Selected Readings in Naval Leadership
for NJROTC Students

Selected Readings in Naval Leadership for NJROTC Students

FIRST EDITION

Edited by Commander Richard R. Hobbs, USNR (Ret.)

Naval Science 4

NAVAL INSTITUTE PRESS • ANNAPOLIS, MARYLAND

Contents

Preface

As a cadet in the NJROTC, you have been exposed to much having to do with the basic qualities of a good follower and an effective leader. It is now time to take a more in-depth look at what leadership is and to learn how to maximize your abilities in this vitally important area.

One of the best methods of polishing leadership skills is to get advice from those who have proven themselves successful leaders. This text is basically a selection of writings on leadership from such individuals, all of whom have been extremely successful in their roles in the naval services of both the United States and other maritime nations. Regardless of these leaders' backgrounds, you will observe many common threads that if followed will lead to success in leadership roles at any level in almost any organization.

In Unit One we revisit the basic theory of leadership and what makes people respond to the leadership effort. Unit Two details the responsibilities of leadership and highlights the qualities that separate outstanding leaders from those not so great. In Unit Three we examine the techniques of effective communication, probably the single most important skill that a good leader must develop in order to be successful.

Unit One contains selected articles on leadership published during the last ten years in the U.S. Naval Institute *Proceedings*, and excerpts from several texts that have long been used in various leadership courses taught in U.S. Navy educational programs. Units Two and Three are excerpts from the unique text, *Naval Leadership: Voices of Experience*, developed over some twenty years by Professor Karel Montor and several other faculty members at the U.S. Naval Academy. It contains thoughts on leadership derived from interviews of over a hundred senior officers of the U.S. Navy and Marine Corps and the naval services of several other maritime nations, as well as observations from other less prominent but nevertheless authoritative individuals such as Master Chief of the Navy (MCPON) Billy C. Sanders, USN (Ret.), whose comments appear at the end of many of the chapters in these two units.

Selections in this book have been edited where necessary to reflect the much-increased role of women as leaders in all echelons of the military services, and indeed most other public and private enterprises of our country.

It is hoped that this text will assist you in adding the polish necessary to be a truly effective leader in your unit, school, and community—and in your chosen occupation.

Selected Readings in Naval Leadership for NJROTC Students

1

Fundamentals of Leadership

Before we can hope to be a success in leading people, we must first be aware of the way in which people are motivated to some course of action. Accordingly, this unit contains the original article on the subject by one of the foremost experts in the field, "A Theory of Human Motivation" by Abraham Maslow. His work, perhaps more than any other on the subject, made leaders and managers in all walks of life aware of the many factors that influence human behavior, and explained seemingly illogical behavior on the part of subordinates in any organization.

Another article far-reaching in its effects, particularly in U.S. Navy circles concerned with leadership and ethical training, is the second chapter in this unit, "Thoughts on Man's Purpose in Life . . . and Other Matters" by Adm. Hyman Rickover. Much of its thought-provoking guidance is as valid today as when it was written some thirty years ago.

The remainder of this unit contains several diverse excerpts from basic Navy leadership texts and manuals, as well as articles on leadership published over the past several years in the U.S. Naval Institute *Proceedings*. All contain excellent guidance for leaders at any level in any organization.

Chapter 1. Leadership Group Dynamics

By Reserve Liaison and Training Branch, Education Center, Marine Corps Development and Education Command

In recent years social scientists have done more and more research in the area of group behavior. This research is aimed at learning how and why groups behave as they do. When individuals are brought together in a group situation, those individual needs take on a new dimension. Each person in the group is no longer influenced only by personal and private needs; he or she is also influenced by the needs and behavior of others in the group. This mixture of needs and behavior in the group creates a complex and dynamic force. Because leadership functions in a group situation (if only a group of two), the leader must have a working knowledge of group dynamics in order to effectively and positively influence the group. In this chapter we discuss the nature of groups, their behavior, and how the leader can use group dynamics to mold an effective group.

There are no set rules or formulas to be applied to a

Source: *Leadership in Action: A Text for U.S. Marine Corps Junior ROTC*. Quantico, Va.: U.S. Marine Corps, 1974.

group of individuals that will immediately result in effective group effort. Not all groups have the same objectives. Neither are they composed of individuals with the same ideas, aptitudes, likes, or dislikes.

Scientific study of the psychological factors of group behavior is relatively new, but already the results of such study are being used by leaders in influencing subordinates in many areas of business, government, the military, and even sports. Such questions are being asked as "What factors cause better teamwork?" and "What makes a mediocre team defeat a much higher-rated team on a given day?" Coaches being interviewed prior to a game are always asked, "Is your team ready for this game?" The question does not refer to the physical strength of the individual members of the team, but to the mental attitude of each member and the team as a whole. Coaches being interviewed following the loss of a game often say that their team was not "up" for the game—that is, they were not sufficiently motivated to give their best effort. Leaders in every field are looking for ways to motivate workers, salespeople, students, athletes, and soldiers. The study of group dynamics is aimed at helping the leader to discover new ways not only to motivate people, but to do it in a way that will provide satisfaction and happiness to those being led.

THE NATURE OF GROUPS

The word "group" when used in reference to people means two or more individuals who have common purposes or characteristics. It implies that the specific members interact and affect each other. Through interaction the members of the group become acquainted with each other, and the group develops a personality of its own. The use of the word "I" gradually changes to "we," and the group starts functioning more as a team with a common purpose.

What actually takes place in this gradual process? True functioning groups are found only when each member gains satisfaction from membership in the group. The use of the word "we" indicates that satisfaction is being gained and the individual is identifying with the group. Thus, in understanding the nature of a group, we must refer to our knowledge of human behavior to find out which individual needs can be satisfied by belonging to a group.

Security through Belonging. It is difficult for an individual to objectively analyze and explain his or her actions. If a man wears his hair long, it might be difficult for him to tell why. He might just like it that way, or perhaps his girlfriend likes it that way. But more than likely, it is because of the influence of the group with which he associates. Group thinking influences social behavior and plays an important role in controlling the general way of life for most of us. It is one of the main controls of social behavior.

Individuals have a desire to be identified with others. This desire could be caused by common interests or goals, or by a person's gregarious nature. Individuals seem to gain a certain amount of security through having a set relationship with others. An individual may belong to several groups at once—the family, a lodge, a fraternity, and a social club. The reason a person belongs to all these organizations is that membership in each group promises to satisfy some needs.

Belonging gives us a feeling of having status, of having a definite purpose or role in the group. It may be a major role or a minor role, but this does not seem to matter. The important thing is that there is a defined place or role in the group. Who is who? What is one's place? How long will the place last? How can one rise in the group? The answers to these questions are important aspects of belonging. They let us know where we stand in relation to those around us. This defined place or role gives us the security of having something stable to work from.

Fulfills the Need for Recognition. All individuals seek status; for this status to be of value it must be recognized by others. Through membership in various groups the individual is seeking a means of gaining some recognition.

The status of an individual is actually a matter of how one feels about one's place in the group. This does not mean that everyone, to have status, wants to be boss, the president, or the supervisor. Status depends on what individuals think about their role or place in the group and on how they define success for themselves in this role. Some individuals prefer a place that is in the limelight, while others prefer the obscure or insignificant role. In other words, each individual has a certain preferred place, status, or role that he or she wants other people to accept and recognize.

An individual may gain feelings of esteem from the status of the group. If the group is well thought of, accomplishes many worthwhile goals, and is looked upon as a group with merit, it brings credit to its members. The Marine Corps has long used this means of satisfying individual esteem needs through the prestige of the group. There is nothing wrong with this method, for it serves as a strong motivating force. Unit citations, honors, and commendations all accomplish the same things, for individuals like to belong to groups they are proud of.

Satisfaction of Other Needs. The belonging and recognition needs are not the only needs that may be satisfied through group activities. The individual may join a group for several reasons. The physician may join a medical society to have access to the latest medical information. The laborer may join a union because he believes his economic status will be improved. The owner of a downtown business may join one of the city's civic organizations in order to have more contact with the public and other businesspeople. Some groups exist solely because individual needs can best be satisfied through united effort. The group that promises satisfaction of more than one need is going to be more attractive than one which offers satisfaction of only a single need.

The group and its activities are very important to the individual in the military service. The change from civilian to military life means learning to live and get along in a new society. The needs that were satisfied through a variety of civilian activities are suddenly blocked because those activities are no longer available. The avenue that is open for the individual to gain some satisfaction is through association with the other members of the group. In military service this association can be much closer than in civilian life, for a military unit may be thrown together with no other contacts for an extended time.

Identification with the Group. This is probably the secret of building successful teams. We know that individuals join or identify with a group because the association promises to pay off in satisfying some deep-rooted needs. These needs must be satisfied by and through the group to cause the individual to wholeheartedly identify with it. On the other hand, the group must benefit from the identification of its members. This means that accomplishing the group's mission is a responsibility of all group members. Another important aspect is that group identification must be satisfying to all members rather than just a selected few. There is no surer way of losing group members and the effectiveness of the group than by favoring a selected few. The group is not a team until *all* members start identifying themselves with it.

Individuals who are effectively identified with the group and its objectives will show this identification by boasting about or showing pride in belonging to the group. Individuals who show this feeling for the group are accepting group success or failure as their own success or failure. On the other hand, some individuals become so strongly identified with the group that they think the group can do no wrong. In this case the individual may become overly dependent on the group and look on those outside as being its enemies. A healthy degree of identification with the group is essential, but not to the extent that the individual loses his or her own identity.

GROUP DYNAMICS

Group dynamics are those forces that result from the interaction of group members among themselves and between the group and the environment in which the group exists. Those forces within the group may be called *internal dynamics.* Those forces arising from interaction of the group with its environment may be called *external dynamics.*

INTERNAL DYNAMICS

The internal dynamics of a group includes those actions resulting from the communication and participation patterns within the group. The behavior and performance of the group is also influenced by such factors as group atmosphere, group standards, group identity, and group spirit.

Communication Patterns. Communication within the group is absolutely essential for effective group functioning. Communication is necessary to provide the information needed to accomplish the group task. Without communication there can be no coordination or cooperation, and there would be no means to develop group attitudes and spirit. Communication is the primary process of group dynamics.

There are many ways by which people in a group communicate. In a formal group there may be established channels of communication, such as the "chain of command" in a military organization. In a democratic group the patterns of communication are more informal. In either case the leader of a group must be at the center of the group communications network. The leader must be in a position not only to direct communications downward to members of the group, but also to receive communication coming up from the group. This two-way communication between the leader and the group is necessary for a number of reasons. First, the two-way exchange of communication enables the leader to determine from the "feedback" from the group whether ideas or instructions have been clearly received and understood by the group. It also provides a means for group members to contribute their ideas and suggestions to the leader. If this upward channel of communication is left open and subordinates are encouraged to use it, many good ideas will originate from the people who are in the best position to recommend changes that will increase efficiency at the working level of the group process. This upward communication also provides a means for subordinates to "let off steam" or to get problems and gripes off their chest. It is not enough, however, to merely leave the door open to upward communication. Leaders must actively encourage and solicit ideas, suggestions, and gripes from members of the group. They can do this through informal discussions with subordinates or by more formal means such as suggestion boxes, or incentives or rewards systems. However they do it, they must create an atmosphere wherein members of the group are encouraged and will feel free to make their ideas known.

Another important communication pattern that exists in all groups is what is known as the "grapevine." This is a term that originated during the Civil War when telegraph wires were often laid in tangled, crisscross patterns that somewhat resembled grapevines. Infantrymen would often tap these telegraph lines in an attempt to get some information about what was going on at the upper echelons of command. This unofficial information was, of course, widely disseminated and was said to come from the "grapevine." The expression today has come to mean any unofficial information that is passed among the members of a group. This information consists of a combination of facts and rumors, but studies of the grapevines of major organizations have shown the information to be highly accurate and reliable. In some cases this information has proven to be 80 to 90 percent accurate. The reason for the reliability of such communication between group members is due in part to the informal atmosphere in which it is transmitted. Since the information is usually passed among friends, it is received in an atmosphere of trust and confidence. This informality helps to avoid some of the psychological blocks that interfere with communication between subordinates and seniors. Because this information is usually something of specific interest to the group, members tend to be receptive to the information and are more likely to remember it and pass it on just as they heard it. Problems arise, however, when the grapevine starts to circulate rumors. Rumors tend to spring up any time there is a lack of factual information about things that are of vital interest to the group. To prevent the spread of rumors, the leader must keep the grapevine supplied with facts. When the group knows the truth about a situation of interest to them, they have no need or desire to spread rumors.

Rather than look at the grapevine as a nuisance, leaders should use it to the advantage of effective group communication. They can learn through the grapevine who the informal leaders of various subgroups are, and through them, leaders can use the grapevine to gain information that might not be available through other means. A leader can also use it to disseminate messages that might be better received in the informal atmosphere of the grapevine.

Participation Patterns. In the group situation it is impossible for every member to participate to the same degree. The varying interests, aptitudes, and talents of the individual members will largely dictate the participation pattern of a particular group. If the group is left to find its own pattern without direction from group leadership, some unfavorable patterns may develop. Some members will tend to overparticipate and dominate the group. Others may fail to participate at all without some stimulus from within the group. In a group discussion, for example, a few members might try to do all the talking, thus depriving others of an opportunity to contribute to the discussion. Frequently a discussion will not bring out what various members might be able to contribute. It is the leader's duty to draw out ideas and opinions from all of the group. The leader must realize that all members have something to contribute, but it is sometimes necessary to encourage participation by those who are shy or reluctant to speak for fear of ridicule.

The leader's participation will also have a considerable influence on the participation pattern of the group. If the leader is autocratic and authoritarian to the degree of making all decisions without seeking the advice and assistance of others, the other members will soon cease to participate in the decision-making process at all. This could also affect their attitude toward the group and its goals. A much better situation is created within the group when the leader draws upon the varied talents that are available in the group. This makes the group aware of its talents, skills, and other resources, and enhances the feelings of worth of individual members.

Group Standards. A common characteristic of all groups is the adoption of group standards. That is the code of operation or behavior that the group adopts in a particular situation. It is a collective sense of responsibility that the group has about how it carries out its work. Sometimes a group's stated standards may not agree with what is the actual practice. For instance, new cadets join the unit and are told that they must maintain a neat appearance, with shined shoes, a proper uniform, and a neat haircut. For a few weeks they come to school that way, but they notice that other cadets have unshined shoes and unpressed uniforms, and nothing is said about it. Soon the new cadets are coming to school with unshined shoes and an otherwise unkempt appearance. They have adapted themselves to the true group standards, which are different from what the new cadets were told when they joined the unit.

It is the responsibility of the leader to help the group to set standards for acceptable performance. It is not enough just to aim for high standards. Group standards should be well defined so that each member knows what is expected. They should also be realistic. If stated standards are so high that they are impossible for the majority of the group to maintain, they will lead only to frustration and dissatisfaction. Group standards should also be consistent and not change at the whim of group leaders, and they should apply to all members of the group. The example that is set by leaders of the group will in the long run have a greater influence on setting the true standards of the group than any formal statement of ideals or aims.

Group Solidarity. Another characteristic of a group is that of group solidarity or cohesion. This is the tendency for members of a group to stick together. This tendency may be very strong, or it may exist only as a fragile bond, but every group exhibits this characteristic to some degree.

Group solidarity is achieved through a variety of motivations. Some groups may stick together for security or protection. If a group is attacked from outside, it comes together more strongly to defend itself from these attacks. An example of this is the story of the Jewish people in Germany under the persecution of Hitler's policy of extermination. As a result of their treatment they developed a solidarity of spirit and determination that had not existed previously among that group.

The goal of the group leader is to develop the kind of group solidarity that will enable members to work together in a cooperative situation in which all members have specific responsibilities and are at the same time interdependent on each other. This kind of solidarity is developed in a number of ways. The first factor in the development of group cohesion is the simple matter of close association. To develop strong emotional ties people have to know each other well, and this can be done only through close association. Close association over a long period of time can develop very strong bonds between group members. The longer a group stays together in an organization, the stronger their feelings of solidarity become. But close association alone is not enough to develop strong bonds among people. It is also necessary for them to have a common purpose. Group purpose and teamwork are essential in building group solidarity.

Another factor that promotes group cohesiveness is common experiences. People who share common experiences develop close identity with each other based on those experiences. These experiences, however, must be satisfying to group members. The sharing of frustrating experiences or failures can break up group solidarity as members leave the group to disassociate themselves from unpleasant experiences. The leader of a group can promote solidarity by providing situations that promote close association through common experiences with a sense of common purpose.

Esprit de Corps. When an organization has a high degree of solidarity and group identity, it is said to have esprit de corps. Esprit de corps has a meaning that transcends the literal translation of the words. It is a quality that is found in formal organizations with a strong sense of history, tradition, and heritage, and which offer its members opportunities for recognition and prestige through membership. It is characterized by enthusiastic pride and loyalty of members, including the display and use of symbols and slogans that are identified with the group. When a group has high esprit, it is often capable of performance that far surpasses what would be normally expected. If this quality is highly developed in an organization, it is possible to pass it on from generation to generation in the same organization even though the entire membership will change over a period of years.

EXTERNAL DYNAMICS

In addition to the internal forces that affect group action and behavior, there are external forces that also have an important influence on the group. These external forces are the factors that affect the group as a whole. Every group exists in some kind of environment. Your Marine Corps Junior ROTC unit exists in the school environment, which in turn is part of a larger community. The attitude of your school and community toward your unit will have an effect on the way your unit performs. Your parent organization, the Marine Corps, is another external force that influences your unit. All of these external groups have a certain image of your unit, and they have certain expectations about how you should perform as a group. The Marine Corps expects you to uphold the same high standards of behavior and performance that are expected of all Marines. The school expects you to represent the ethics, standards, and values of the school.

Sometimes there may be conflicts between the expectations of some of these outside influences. For example, the standards of dress, haircuts, and appearances may differ between what the school population expects and what your Marine Instructors expect. Such influence will have an effect on the behavior of the unit and must be considered in understanding its functioning.

Another factor of external group dynamics is the membership of group members in other groups. Members of a group may belong to a number of groups, all of which are competing for the individual's time and energy. Since time is a limited commodity, the individual must make a choice of which group is most important. Usually the individual will choose the group that offers the greatest opportunity to achieve security, acceptance, recognition, and the other basic needs.

No group exists in a vacuum, and group behavior cannot be considered without reference to external forces. The group leader must be aware of these external forces and see that group goals are kept in harmony with the goals of the community in which the group exists.

THE GROUP LEADER

The overall objective of a leader is to motivate the group toward the achievement of its mission or goals. To do this the leader must create within the group a satisfying social structure. There are a number of techniques that can be used to achieve this kind of group situation.

Identify Goals. The leader must help the group members to identify specific goals so that they will have a clear picture of what they are about to accomplish. If it is a democratic group with goals that are derived by consensus, it may be necessary for the group leader to distill some concrete and well-defined goals out of the generalized objectives of the group. If it is an autocratic group with goals or missions assigned by higher authority, the leader must define those goals or missions so that they may be clearly understood by the group and in such a way that the group members can identify with the goals and accept them as their own. In either case the achievement of group goals must provide for satisfaction of needs of individual group members. Once the leader has identified the group's goals, the task may be more than half done. Once these goals are defined, the internal forces and energies of a responsive group will move toward their accomplishment.

Give Direction. The leader must give direction to the group effort. A planned procedure is essential for efficient group functioning. This involves management and means establishing an effective and satisfying framework for group action. The plan must be meaningfully communicated to all members for best results. This means letting everyone in on the plan, for individuals need to feel that they are involved in and are an important part of the plan. Even greater satisfaction may be achieved if group members are allowed to participate in the actual planning itself. This participation in the planning enables individual members to select suitable and satisfying roles for themselves and leads to closer identification with the group goals.

Give Consistent Treatment. Consistent treatment of group members creates a secure environment for group members. This means being just and giving to others what they are due in your honest opinion. This is especially important in an autocratic organization like the military. When subordinates know the standard of behavior that is expected of them and the respective rewards and punishments, they will want to abide by that standard, especially when it applies to all members alike. They will feel more secure if they know that their rewards and punishments are awarded objectively, based on their behavior, rather than on the leader's whims, moods, or pet preferences. On the other hand, if their treatment is inconsistent, the group members do not know how to behave, and they question their security. Subordinates do not like for the group to be punished for the behavior of the few. It doesn't make sense to upset the whole group to punish a few offenders. The leader who is consistent in his or her treatment of subordinates will have a minimum of discipline problems.

Create Status. Creating status for the group as

well as for the individual is necessary for effective leadership. Recognized successes add great prestige to any group. It makes group members feel proud of their affiliation, and they show this pride in their behavior. All jobs or roles in the group should be considered important, and all group members should be made to see how their efforts, through the group, help to achieve the goals of the group.

As a cadet leader you will have a certain status in your unit. You will be working with cadets of various ranks. Each individual, no matter what rank or grade, deserves status in the group. No one wants to be looked upon as a nobody, as unimportant, or as a number. You can help individual cadets to gain and maintain their status by giving recognition for accomplishments, looking after their welfare, and showing genuine interest.

The long-recommended technique of "praise in public, reprimand in private" still holds true. Public praise gives an individual status, while public reprimand takes away status. However, praise should be deserved and warranted. Undeserved praise may be worse than no praise at all. A good rule to follow is: When praise comes from without, pass it on to the group or individual, minimizing your part. When the group or a subordinate is censured, accept it as your own responsibility. Shoulder the blame for any group failure. In all military services the leader of a military unit is responsible for everything the unit does or fails to do.

SUMMARY

Scientific study of group behavior is a relatively new field, but results from such studies are being used by present-day leaders. The word "group" refers to two or more individuals who have common purposes or characteristics. The behavior of the group is influenced by the interests of its members.

In studying the nature of the group we need to find out which individual needs can be satisfied by belonging to a group. Individuals have a desire to be identified with others, and they like to belong to groups through which they can gain feelings of security, recognition, and status.

There are many internal and external forces working in a group. Some of these forces may work against leadership. It is the leader's responsibility to influence these forces to work in achieving the goals of the group.

Getting all members to identify with the group and its objectives is probably the secret of building successful teams. This identification with the group must show promise of paying off by satisfying some of the needs of the individual. The more needs satisfied by and through the group, the stronger the identification will be.

To make group dynamics work in achieving group goals, the leader must create a satisfying social structure within the group. The leader can promote this kind of motivation by identifying goals and giving direction to group activities. The leader must also give consistent treatment to all group members alike, and create status for individuals and the group as a whole.

Chapter 1. Study Guide Questions

1. What influences people in a group?
2. What membership condition characterizes true functioning groups?
3. What are some of the effects of a group on individuals within the group?
4. Upon what does an individual's perceived status within a group depend?
5. What must a leader do to guard against losing group members and lessening the effectiveness of the group?
6. What are the main factors that influence the internal dynamics of a group?
7. What formal and informal factors regarding communications within a group must an effective leader keep in mind?
8. What will largely determine the participation patterns of a particular group?
9. What are group standards?
10. A. What is group solidarity?
    B. How is it achieved?
11. What characterizes an organization with high esprit de corps?
12. What are the external group dynamics factors that can influence a group?
13. What are four things a leader can do to motivate group members toward the achievement of their mission or goals?
14. What is a long-recommended technique for giving praise and reprimands?

Vocabulary

new dimension	external dynamics
psychological factors	the "grapevine"
status	prestige
group dynamics	autocratic
internal dynamics	

Chapter 2. Thoughts on Man's Purpose in Life . . . and Other Matters

By Adm. Hyman G. Rickover

Voltaire once said, "Not to be occupied and not to exist are one and the same thing for a man." With those few words Voltaire captured the essence of a purpose to life: to work, to create, to excel, and to be concerned about the world and its affairs.

"We measure ourselves by many standards," said William James, nearly a century ago. "Our strength and our intelligence, our wealth and even our good luck, are things which warm our heart and make us feel ourselves a match for life. But deeper than all such things, and able to suffice unto itself without them, is the sense of the amount of effort which we can put forth." Humans have a large capacity for effort. But it is so much greater than we think it is, few ever reach this capacity.

Work begins with a job, or profession. Having a vocation is always somewhat of a miracle, like falling in love. I can understand why Martin Luther said that a man is justified by his vocation, for it is already a proof of God's favor. But having a vocation means more than punching a timeclock. One must guard against banality, ineptitude, incompetence, and mediocrity.

Source: Cdr. John B. Washbush, USNR, and LCdr. Barbara J. Sherlock, USN, eds. *To Get the Job Done.* Annapolis: Naval Institute Press, 1981.

We as a people seem inclined to accept average or mediocre performance. Mediocrity can destroy us just as surely as perils far more famous. It is important that we remember to distinguish between what it means to fail at a task and what it means to be mediocre. There is all the difference in the world between the life lived with dignity and style which ends up failing, and one which achieves power and glory yet is dull, unoriginal, unreflective, and mediocre. In a real sense, what matters is not so much whether we make a lot of money, hold a prestigious job, or whether we don't; what matters is that we become people who seek out others with knowledge and enthusiasm—that we become people who can enjoy our own company. In the end, learning to avoid mediocrity gives all of us the chance to discover that success really comes in making ourselves into educated individuals, able to recognize that there is always a difference between living with excellence and living with mediocrity. Sherlock Holmes once told Dr. Watson, "Watson, mediocrity knows nothing higher than itself. It takes talent to recognize genius." To which he could have added, it takes talent to know that what counts is condemning mediocrity not in others but in ourselves.

It is a device of the devil to let sloth into the world. One cannot work at a profession alone without stagnating: some time away from the job can be as much or more important in one's overall development. Intellectually, we must never stop growing.

Our conscience should never release us from concern for the problems of the day. Our minds must be forever skeptical yet questioning. We must strive to be singularly free from that failing so common to humankind, deplored by Pascal in the *Pensées*, of filling our leisure with meaningless distractions so as to preclude the necessity of thought. To be an intellectual in the fullest sense, one's mind must be in constant movement.

Intelligence—the central virtue of moral life—is being able to judge the limitations of knowledge. Though there is no substitute for intelligence, it is not enough. People may be intelligent but lack the courage to act. To find a purpose to life, one must be willing to act, to put excellence in one's work and concern for what is right before personal safety.

No professional person has the right to prefer personal peace to the happiness of humankind; one's place and one's duty are to be in the frontline of those who struggle, not in the unperturbed ranks of those who keep themselves aloof from life. If a profession is to have its proper place in the further development of society, it must be increasingly dissatisfied with things as they are. If there is to be any exaltation in our work, we must learn to reach out, not to struggle for that which is just beyond, but to grasp at results which seems almost infinite. As Robert Browning wrote, "Ah, but a man's reach should exceed his grasp, or what's a Heaven for."

The deepest joy in life is to be creative. To find an undeveloped situation, to see the possibilities, to decide upon a course of action, and then devote the whole of one's resources to carrying it out, even if it means battling against the stream of contemporary opinion, is a satisfaction in comparison with which superficial pleasures are trivial. But to create, you must care. You must be willing to speak out.

A certain measure of courage in the private citizen is necessary for the good conduct of the state. Otherwise people who have power through riches, intrigue, or office will administer the state at will and ultimately to their private advantage. For the citizen, this courage means a frank exposition of a problem and a decrying of the excesses of power. It takes courage to do this, for in our polite society frank speech is discouraged. But when this attitude relates to questions involving the welfare or survival of the nation, it is singularly unfitting to remain evasive. It is not only possible, but in fact the duty of everyone to state precisely that which knowledge and conscience compel them to say. Only complete candor and frankness, deep respect for the facts, however unpleasant and uncomfortable, great efforts to know them where they are not readily available, and drawing conclusions guided only by rigorous logic can bring many of today's problems forward.

Jules Cambon said, "We have to defend the country against mediocrity: Mediocrity of soul, mediocrity of ideas, mediocrity of action. We must also fight against it in ourselves." It is a lonely and hard thing to speak out against mediocrity or the evils of the system. One feels the frustration which arises only in those who are compelled to act. The detached spectators do not feel this helplessness because they never try to surmount the problems that exist.

Spectators are those who have no conscience, who don't die, who cannot laugh, and who are unaware of personal responsibility. They do not necessarily do wicked things, but they do passive things. There are many spectators in the world today who choose passivity. They will do anything so that their accustomed life is not disturbed, anything so as not to have to cross over into hardship today; at the same time, they hope tomorrow will take care of itself.

DECLINING RESPONSIBILITY

Most of the work in the world today is done by those who work too hard; they comprise a "nucleus of martyrs." The greater part of the remaining workers' energy goes into complaining. Employees today seldom become emotional about their organization or its output; they are only interested in getting ahead. And many organizations are killing their employees with kindness, undercutting their sense of responsibility with an ever-increasing permissiveness. This is a fatal error, for where responsibility ends, performance ends also.

The sense of responsibility for doing a job right seems to be declining. In fact, the phrase "I am not responsible" has become a somewhat standard response in our society to complaints of a breakdown in the system. This response is a semantic error. Generally what a person means is: "I cannot be held legally liable." Yet, from a moral or ethical point of view the statement is quite true. The person or organization taking this way out is truly not responsible; the person or organization is *irresponsible*.

The unwillingness to act and to accept responsibility is a symptom of America's growing self-satisfaction with the status quo. The result is a paralysis of the spirit, entirely uncharacteristic of Americans during the previous stages of their history. Even the complaints about high taxes and high prices are illusory. Behind them is hidden the reality that the majority, in terms of sheer creature comfort, never had it so good. Those who are still on the outside looking in are not strong or numerous enough to make a political difference.

A major reason why so large a majority is smugly docile is that it has accepted the unwritten rules of the game: Don't rock the boat as long as you get your cut.

Why become worked up over corruption as long as there are enough benefits of the fallout to go around? Once the acceptance of corruption becomes sufficiently widespread, effective exposure seems threatening to too many people and interests. Clamor for closing loopholes declines in direct proportion to the number of people who benefit from loopholes of their own. Freedom of speech seems less important when the majority persuades itself that it is not likely ever to want to speak out to complain. . . .

We shall not get through our troubles safely until a considerable number of Americans acknowledge that they themselves are a part of the process by which the society will be made whole.

The crucial element of hope amid the current drift lies in the recollection of what the American spirit of activist optimism has so often accomplished—when it was rallied by its leaders and philosophers to a vision of shared goals.

For the person who strives to excel, to shoulder responsibility and to speak out, there are enemies in every direction. The enemies are those who have a total willingness to delegate worries about the world to officialdom. They assume that only the people in authority are in a position to know and act. They believe that if vital information essential to the making of public decisions is withheld, it can only be for a good reason. If a problem is wholly or partially scientific in nature, they will ask no questions even though the consequences of the problem are political or social.

The enemies are those whose only concern about the world is that it stay in one piece during their own lifetime. They are invariably successful and regard their good fortune not as a challenge to get close to the real problems of the age, but as proof of the correctness of everything they do. Nothing to them is less important than the shape of things to come or the needs of the next generation. Talk of the legacy of the past or of human destiny leaves them cold. Historically, they are the disconnected. They are enemies because they are detached from the kind of concern for the rights of unborn legions that will enable the world itself to become connected and whole.

To struggle against these enemies, and against apathy and mediocrity, is to find the purpose in life.

EDUCATION

There are five basic goals for education. When I say "basic," I mean goals that apply to everyone—men or women, rich or poor.

Goal 1: To learn to understand, appreciate, and take care of the natural world we live in. Most people go through life unaware of the fascinating complex of events around them, of climate and terrain and vegetation and animals and people and their interrelatedness. Civilized people need to know not only what the environment is like, but how to keep it habitable.

Goal 2: To understand, appreciate, and learn to live with the fellow inhabitants of our planet. All children must learn about the races and people of the world and the rich variety of the world's cultures. They must know something of the history of people and of nations. They must learn that there are many people in the world who differ from them profoundly in habits, ideas, and ways of life. They must perceive these differences not as occasions for uneasiness or hostility but as challenges to their capacity for understanding.

Goal 3: To have an area of esthetic experience— and I would include the religious and spiritual with the esthetic. The esthetic experience is the organization of our feelings—the search for and the creation of orders in our affective life. The significance, the meanings that we perceive, are private. To give ourselves, for at least a part of the time, to the lonely contemplation of some kind of beauty and order is also to enrich ourselves so that we have something to contribute to the lives of others.

Goal 4: To earn a living. This can be learned in school or out, and at any level from humble work to highly paid, professional skills.

Each of us needs to feel, sometime in life, that our services are important enough so that someone other than the welfare department is willing to pay to keep us alive. Those who have never proved their usefulness remain forever at a disadvantage, because work is a basic way in which most of us relate to the world.

But work is, in a profound way, held in contempt by our educational system. Students believed to be low in academic talent are steered into vocational programs, while academically more gifted students are steered away from them, as if they were too good to be made to work. Such a distinction is arbitrary and invidious and, just as the rain falls equally on the just and the unjust, it inflicts an injustice both on the academically slow and on the advanced. Throughout all our high schools and colleges, there should be maintained an active relationship between the academic world and the world where people earn a living.

Goal 5: To learn some kind of critical or intellectual method. I regard this as the most important of all. We have all learned that we live in the age of an information explosion. But we are also in the middle of a misinformation explosion. With the proliferation of mass communications media, we are surrounded by hawkers, hard and soft sells, persuaders hidden and overt. Bombarded daily with millions of words by print and electronic media, we all have to have some kind of critical method by means of which to decide whom and what to believe, and to what degree.

How is propaganda evaluated? It cannot always be analyzed by scientific method, since propagandistic statements are rarely capable of proof, but it can be approached with a scientific attitude. Some kind of discipline in the orientations of science is necessary to inculcate a critical attitude toward words, our own as well as those of others, so that our lives may be governed by the skepticism and respect for fact that characterize the rational mind.

IMPORTANCE OF READING

The proof of living, as Norman Cousins has said, "is in memory, and all of us, through reading, can live five or six lifetimes in one. Through reading, the sluices of the mind open up, making accessible a range of experiences otherwise beyond our personal reach." In reading books, we grow both emotionally and intellectually. Marya Mannes feels "a person can be fascinated by movies and diverted by television, but they are a series of snacks. Books are the real nourishment of the human mind."

It has been said of books that they "cause us to reflect on our own perceptions, make distinctions of new areas of reality and become sensitive to them. It is as our teachers used to say—through reading you start thinking about the world in new ways." It has also been said that "the essence of a classic is that it arises from such profound depths in human experience that it speaks to us, who live centuries later in vastly different cultures as the voice of our own experience; it helps us to understand ourselves better and enriches us by releasing echoes within ourselves which we may not have known were there."

When we read a truly great book, we are lost in admiration. A great book is so powerful that it forces us immediately to reread it, because we have been altered in the process of reading it.

Thomas Babington Macaulay said of great literature:

These are the old friends who are never seen with new faces, who are the same in wealth and in poverty, in glory and in obscurity. With the dead there is no rivalry. In the dead there is no change. Plato is never sullen. Cervantes is never petulant. Demosthenes never comes unseasonably. Dante never stays too long. No difference of political opinion can alienate Cicero. No heresy can excite the horror of Bossuet.

It is crucial that we inspire the young to read; for if they do not read, they will never change this world or beat its manipulators. They might start with the Federalist Papers and work through Abraham Lincoln and Winston Churchill to learn not only the art of persuasion but the content of integrity.

To be effective in the world, it is not sufficient to be well read. One must also write clearly and succinctly. It is not easy to write well. It means studying to gain a good vocabulary, and practicing to learn how to use it. Joseph Conrad compared writing to carrying heavy bales under a low rope on a hot day. It has been said that good writing is inherently subversive; I agree. Good writing can last for years, stimulating and inspiring those who read it, and causing them to reassess their views of the world and its institutions. But then, of course, we have gone full circle to the importance of reading books.

THE PAST

We need to regain some of the certainties of the past. For the uneasiness and the malaise of our time is due to a root cause: In our politics and economy, in family life and religion, in practically every sphere of our existence the certainties of the eighteenth and nineteenth centuries have disintegrated or been destroyed. Much of the social cement that has held our society together—shared values, strong family structure, the influence of the church and the local community—has been dissolving steadily over the years. At the same time, no new sanctions or justifications for the new routines we live, and must live, have taken hold. So there is no acceptance and there is no rejection, no sweeping hope and no sweeping rebellion. There is no real plan of life.

Many certainties of the past have been lost through ill-founded criticisms of past customs and institutions. It is easy in the light of present-day knowledge and achievement to ridicule and condemn the people, the ideas, and the customs of a past age. But this is a mistake. Much of the wisdom and most of the beauty of the past is lost because we do not have sufficient knowledge and imagination to divest ourselves of our modern refinements and to relive the struggles, the hopes and fears, and the faiths and beliefs of our ancestors who were in contact, like us, with the mystery of life.

I would urge a more sympathetic understanding of the past, so that we may recapture some of its beauty. Certainly, some of the past is not worthy of preservation. But much is. In nature, the demolition process is a living process; the destruction of the old is caused by and is part of the generation of new life. In all areas of human existence, the present is the fruit of the past and the seed of the future.

RELIGION

Religion is one area that merits a more sympathetic understanding by modern intellectuals. For religion has not so much declined as it has changed. It is true that religion no longer has direct influence on the large corporate structures that have emerged in the past 400 years, such as government, business, labor, the military, and education. It has had to relinquish many of its "mysteries" to the explanations of rational science. Our increased capacity for abstract thought has made a similar penetration of religion's myths, which are now treated more as metaphors than absolutes. And, finally, the church doors have been thrown open to free choice, to more explicit or individual formulations than previously.

What has not changed are the basic functions of religion: Supplying a faith or meaning for coping with the questions of the human condition; providing for a feeling of belonging, in the broadest sense.

Some sociologists are fond of describing these functions today as "residual" of "inauthentic." To these sociologists, all primordial or prerationalities are considered unenlightened and reactionary. If people are religious, they are not to be trusted; more, they are to be mistrusted, as people living in a superstitious and narrow-minded backwater of social evolution. This seems to be a rather arrogant attitude, especially considering that through the influence of modern civilization, people appear to have "progressed" from physical to spiritual starvation.

If we are to regain some of the certainties of our life, we must understand and incorporate the most universal and worthy ideas of our past into our present existence.

Woodrow Wilson expressed this well in his last writing before he died. He warned the American people that their society, unless it was redeemed by being permeated with the spirit of Christ, could not survive. This meant a social and economic order based on "sympathy and helpfulness and a willingness to forgo self-interest in order to promote the welfare, happiness, and contentment of others and of the community as a whole."

It makes little difference what particular religion a person follows. The important thing is to live up to its precepts. More and more as I grow older, I comprehend there are certain basics in life to which a person must adhere.

MORALS AND ETHICS

There is abundant evidence around us for one to conclude that morals and ethics are becoming less prevalent in people's lives. The standards of conduct which lay deeply buried in accepted thought for centuries are no longer absolute. People seem unable to differentiate between physical relief and moral satisfaction. Many confuse material success in life with virtue.

The decline in morals parallels the decline of traditional religion in all areas of our society. In our desire to separate church and state, we have gone to the opposite extreme and have excised religious training from our public schools and colleges, thus depriving our youth of the lasting standards of morals and ethics as enunciated by the Ten Commandments and the Sermon on the Mount.

Morals are the quarrel we have with behavior. Yet, any system of education which does not inculcate moral values simply furnishes the intellectual equipment whereby men and women can better satisfy their pride, greed, and lust.

We are now living on the accumulated moral capital of traditional religion. It is running out, and we have no other consensus of values to take its place. This is partly so because we can now obtain on earth what previously was promised us when we reached heaven.

In our system of society, no authority exists to tell us what is good and desirable. We are each free to seek what we think is good in our own way.

We should all think about what is the basis for a consensus of values in our society. Where do our goals and values come from? If we believe in a free society, what limitations, if any, are we willing to place on that freedom so that society may protect and maintain itself?

Chapter 2. Study Guide Questions

1. What does Admiral Rickover say about the human capacity for effort?

2. What must one do to find a purpose in life?

3. According to Admiral Rickover, what is necessary to solve many of today's problems?

4. A. What indicates that more and more Americans are satisfied with the status quo of our society?

   B. What is the result?

5. What are the five basic goals for education?

6. Why is reading important?

7. Why should we seek to understand the past?

Vocabulary

vocation	subversive
banality	metaphor
ineptitude	sociologist
incompetence	status quo
mediocrity	apathy
sloth	esthetic
skeptical	clinical
exaltation	paralysis
decry	illusory
intrigue	docile
corruption	malaise
invidious	primordial
propaganda	

Chapter 3. A Theory of Human Motivation

By Abraham H. Maslow

Any theory of human motivation that could lay claim to being definitive would include the following propositions.

1. The integrated wholeness of the organism must be one of the foundation stones of motivation theory.

2. The hunger drive (or any other physiological drive) is rejected as a centering point or model for a definitive theory of motivation. Any drive that is somatically based and localizable is atypical rather than typical in human motivation.

3. Such a theory should stress and center itself upon ultimate or basic goals rather than partial or superficial ones, upon ends rather than means to these ends. Such a stress would imply a more central place for unconscious than for conscious motivations.

4. There are usually available various cultural paths to the same goal. Therefore conscious, specific, local-cultural desires are not as fundamental in motivation theory as the more basic, unconscious goals.

5. Any motivated behavior, either preparatory or consummatory, must be understood to be a channel through which many basic needs may be simultaneously expressed or satisfied. Typically an act has *more* than one motivation.

6. Practically all organismic states are to be understood as motivated and as motivating.

7. Human needs arrange themselves in hierarchies of prepotency. That is to say, the appearance of one need usually rests on the prior satisfaction of another, more prepotent need. Human beings are perpetually wanting animals. Also no need or drive can be treated as if it were isolated or discrete; every drive is related to the state of satisfaction or dissatisfaction of other drives.

8. *Lists* of drives will get us nowhere for various theoretical and practical reasons. Furthermore any classification of motivation must deal with the problem of levels of specificity or generalization of the motives to be classified.

9. Classifications of motivations must be based upon goals rather than upon instigating drives or motivated behavior.

10. Motivation theory should be human-centered rather than animal-centered.

11. The situation or the field in which the organism reacts must be taken into account but the field alone can rarely serve as an exclusive explanation for behavior. Furthermore the field itself must be interpreted in terms of the organism. Field theory cannot be a substitute for motivation theory.

Source: *Psychological Review*, vol. 50 (1943), pp. 370–96.

12. Not only the integration of the organism must be taken into account, but also the possibility of isolated, specific, partial, or segmental reactions.

13. Motivation theory is not synonymous with behavior theory. The motivations are only one class of determinants of behavior. While behavior is almost always motivated, it is also almost always biologically, culturally, and situationally determined as well.

The present paper is an attempt to formulate a positive theory of motivation which will satisfy these theoretical demands and at the same time conform to the known facts, clinical and observational as well as experimental. It derives most directly, however, from clinical experience. This theory is, I think, in the functionalist tradition of William James and John Dewey, and is fused with the holism of Max Wertheimer, Kurt Goldstein, and gestalt psychology, and the dynamicism of Sigmund Freud and Alfred Adler. This fusion or synthesis may arbitrarily be called a "general-dynamic" theory.

It is far easier to perceive and to criticize the aspects in motivation theory than to remedy them. Mostly this is because of the very serious lack of sound data in this area. I conceive this lack of sound facts to be due primarily to the absence of a valid theory of motivation. The present theory then must be considered to be a suggested program or framework for future research and must stand or fall, not so much on facts available or evidence presented, as on researches yet to be done, researches suggested perhaps by the questions raised in this paper.

THE BASIC NEEDS

THE "PHYSIOLOGICAL" NEEDS

The needs that are usually taken as the starting point for motivation theory are the so-called physiological drives. Two recent lines of research make it necessary to revise our customary notions about these needs, first, the development of the concept of homeostasis, and second, the finding that appetites (preferential choices among foods) are a fairly efficient indication of actual needs or lacks in the body.

Homeostasis refers to the body's automatic efforts to maintain a constant, normal state of the bloodstream. Walter Cannon has described this process for (1) the water content of the blood, (2) salt content, (3) sugar content, (4) protein content, (5) fat content, (6) calcium content, (7) oxygen content, (8) constant hydrogen-ion level (acid-base balance) and (9) constant temperature of the blood. Obviously this list can be extended to include other minerals, the hormones, vitamins, and so forth.

P. T. Young has summarized the work on appetite in its relation to body needs. If the body lacks some chemical, the individual will tend to develop a specific appetite or partial hunger for that food element.

Thus it seems impossible as well as useless to make any list of fundamental physiological needs, for they can come to almost any number one might wish, depending on the degree of specificity of description. We cannot identify all physiological needs as homeostatic. That sexual desire, sleepiness, sheer activity, and maternal behavior in animals are homeostatic has not yet been demonstrated. Furthermore, this list would not include the various sensory pleasures (tastes, smells, tickling, stroking), which are probably physiological and which may become the goals of motivated behavior.

These physiological drives or needs are to be considered unusual rather than typical because they are isolable, and because they are localizable somatically. That is to say, they are relatively independent of each other, of other motivations, and of the organism as a whole, and second, in many cases, it is possible to demonstrate a localized, underlying somatic base for the drive. This is true less generally than has been thought (exceptions are fatigue, sleepiness, maternal responses) but it is still true in the classic instances of hunger, sex, and thirst.

It should be pointed out again that any of the physiological needs and the consummatory behavior involved with them serve as channels for all sorts of other needs as well. That is to say, the person who thinks he is hungry may actually be seeking more for comfort, or dependence, than for vitamins or proteins. Conversely, it is possible to satisfy the hunger need in part by other activities such as drinking water or smoking cigarettes. In other words, relatively isolable as those physiological needs are, they are not completely so.

Undoubtedly these physiological needs are the most prepotent of all needs. What this means specifically is that in the human being who is missing everything in life in an extreme fashion, it is most likely that the major motivation would be the physiological needs rather than any others. A person who is lacking food, safety, love, and esteem would most probably hunger for food more strongly than for anything else.

If all the needs are unsatisfied, and the organism is then dominated by the physiological needs, all other needs may become simply nonexistent or be pushed into the background. It is then fair to characterize the whole organism by saying simply that it is hungry, for consciousness is almost completely preempted by hunger. All capacities are put into the service of hunger-satisfaction, and the organization of these capacities is almost entirely determined by the one purpose of satisfying hunger. The receptors and effectors, the intelligence, memory habits, all may now be defined simply as hunger-gratifying tools. Capacities that are not useful for this purpose lie dormant, or are pushed into the background. The urge to write poetry, the desire to acquire an automobile, the interest in American history, the desire for a new pair of shoes are, in the extreme case, forgotten or become of secondary importance. For the man who is extremely and dangerously hungry, no other interests exist but food. He dreams food, he remembers food, he thinks about food, he emotes only about food, he perceives only food, and he wants only food. The more subtle determinants that ordinarily fuse with the physiological drives in organizing even feeding, drinking, or sexual behavior may now be so completely overwhelmed as to allow us to speak at this time (but *only* at this time) of pure hunger drive and behavior, with the one qualified aim of relief.

Another peculiar characteristic of the human organism when it is dominated by a certain need is that the whole philosophy of the future tends also to change. For our chronically and extremely hungry man, Utopia can be defined very simply as a place where there is plenty of food. He tends to think that if only he is guaranteed food for the rest of his life, he will be perfectly happy and will never want anything more. Life itself tends to be defined in terms of eating. Anything else will be defined as unimportant. Freedom, love, community feeling, respect, philosophy, may all be waved aside as fripperies that are useless since they fail to fill the stomach. Such a man may fairly be said to live by bread alone.

It cannot possibly be denied that such things are true but their *generality* can be denied. Emergency conditions are, almost by definition, rare in the normally functioning peaceful society. That this truism can be forgotten is due mainly to two reasons. First, rats have few motivations other than physiological ones, and since so much of the research on motivation has been made with these animals, it is easy to carry the rat picture over to the human being. Second, it is too often not realized that culture itself is an adaptive tool, one of whose main functions is to make the physiological emergencies come less and less often. In most of the known societies, chronic extreme hunger of the emergency type is rare, rather than common. In any case, this is still true in the United States. The average American citizen is experiencing appetite rather than hunger when he or she says "I am hungry." Sheer life-and-death hunger is apt to be experienced only by accident and then only a few times through an entire life.

Obviously a good way to obscure the "higher" motivations, and to get a lopsided view of human capacities and human nature, is to make the organism extremely and chronically hungry or thirsty. Anyone who attempts to make an emergency picture into a typical one, and who will measure all human goals and desires by behavior during extreme physiological deprivation is certainly being blind to many things. It is quite true that man lives by bread alone—when there is no bread. But what happens to desires when there *is* plenty of bread and when one's belly is chronically filled?

At once other (and "higher") needs emerge and these, rather than physiological hungers, dominate the organism. And when these in turn are satisfied, again new (and still "higher") needs emerge, and so on. This is what we mean by saying that the basic human needs are organized into a hierarchy of relative prepotency.

One main implication of this phrasing is that gratification becomes as important a concept as deprivation in motivation theory, for it releases the organism from the domination of a relatively more physiological need, permitting thereby the emergence of other more social goals. The physiological needs, along with their partial goals, when chronically gratified cease to exist as active determinants or organizers of behavior. They now exist in only a potential fashion in the sense that they may emerge again to dominate the organism if they are thwarted. But a want that is satisfied is no longer a want. The organism is dominated and its behavior organized only by unsatisfied needs. If hunger is satisfied, it becomes unimportant in the current dynamics of the individual.

This statement is somewhat qualified by a hypothesis to be discussed more fully later, namely, that it is precisely those individuals in whom a certain need has always been satisfied who are best equipped to tolerate deprivation of that need in the future, and that furthermore, those who have been deprived in the past will react differently to current satisfactions than the one who has never been deprived.

THE SAFETY NEEDS

If the physiological needs are relatively well gratified, there then emerges a new set of needs, which we may categorize roughly as the safety needs. All that has been said of the physiological needs is equally true, although in lesser degree, of these desires. The organism may equally well be wholly dominated by them. They may serve as the almost exclusive organizers of behavior, recruiting all the capacities of the organism in their service, and we may then fairly describe the whole organism as a safety-seeking mechanism. Again we may say of the receptors, the effectors, of the intellect and the other capacities that they are primarily safety-seeking

tools. Again, as in the hungry man, we find that the dominating goal is a strong determinant not only of his current world outlook and philosophy but also of his philosophy of the future. Practically everything looks less important than safety (even sometimes the physiological needs which, being satisfied, are now underestimated). A man, in this state, if it is extreme enough and chronic enough, may be characterized as living almost for safety alone.

Although in this paper we are interested primarily in the needs of adults, we can approach an understanding of their safety needs perhaps more efficiently by observation of infants and children, in whom these needs are much more simple and obvious. One reason for the clearer appearance of the threat or danger reaction in infants is that they do not inhibit this reaction at all, whereas adults in our society have been taught to inhibit it at all costs. Thus even when adults do feel their safety to be threatened, we may not be able to see this on the surface. Infants will react in a total fashion and as if they were endangered, if they are disturbed or dropped suddenly, startled by loud noises, flashing light, or other unusual sensory stimulation, by rough handling, by general loss of support in the mother's arms, or by inadequate support. As the child grows up, sheer knowledge and familiarity as well as better motor development make these "dangers" less and less dangerous and more and more manageable. Throughout life it may be said that one of the main conative functions of education is this neutralizing of apparent dangers through knowledge, for example, I am not afraid of thunder because I know something about it.

In infants we can also see a much more direct reaction to bodily illnesses of various kinds. Sometimes these illnesses seem to be immediately and per se threatening and seem to make the child feel unsafe. For instance, vomiting, colic, or other sharp pains seem to make the child look at the whole world in a different way. At such a moment of pain, it may be postulated that, for the child, the appearance of the world suddenly changes from sunniness to darkness, so to speak, and becomes a place in which anything at all might happen, in which previously stable things have suddenly become unstable. Thus a child who because of some bad food is taken ill may, for a day or two, develop fear, nightmares, and a need for protection and reassurance never seen before the illness.

Another indication of the child's need for safety is a preference for some kind of undisrupted routine or rhythm. The child seems to want a predictable, orderly world. For instance, injustice, unfairness, or inconsistency in the parents seems to make a child feel anxious and unsafe. This attitude may be not so much because of the injustice per se or any particular pain involved, but rather because this treatment threatens to make the world look unreliable, or unsafe, or unpredictable. Young children seem to thrive better under a system which has at least a skeletal outline of rigidity, in which there is a schedule of a kind, some sort of routine, something that can be counted upon, not only for the present but also far into the future. Perhaps one could express this more accurately by saying that the child needs an organized world rather than an unorganized or unstructured one.

The central role of the parents and the normal family setup is indisputable. Quarreling, physical assault, separation, divorce, or death within the family may be particularly terrifying. Also, parental outbursts of rage or threats of punishment directed to the child, name calling, speaking harshly, shaking, rough handling, or actual physical punishment sometimes elicit such total panic and terror in the child that we must assume more is involved than the physical pain alone. While it is true that in some children this terror may represent a fear of loss of parental love, it can also occur in completely rejected children, who seem to cling to the hating parent more for sheer safety and protection than because of hope of love.

Confronting the average child with new, unfamiliar, strange, unmanageable stimuli or situations will too frequently elicit the danger or terror reaction, as for example, getting lost or even being separated from the parent for a short time; being confronted with new faces, new situations, or new tasks; the sight of strange, unfamiliar, or uncontrollable objects; illness; or death. Particularly at such times, the child's frantic clinging to the parent is eloquent testimony to the parent's role as protector (quite apart from the roles as food-giver and love-giver).

From these and similar observations, we may generalize and say that the average child in our society generally prefers a safe, orderly, predictable, organized world, which can be counted on, and in which unexpected, unmanageable, or other dangerous things do not happen, and in which, in any case, the all-powerful parent will protect and shield the child from harm.

That these reactions may so easily be observed in children is in a way a proof of the fact that children in our society feel too unsafe (or, in a word, are badly brought up). Children who are reared in an unthreatening, loving family do not ordinarily react as we have described above. In such children the danger reactions are apt to come mostly to objects or situations that adults too would consider dangerous.

The healthy, normal, fortunate adult in our culture is largely satisfied in the safety needs. The peaceful, smoothly running, "good" society ordinarily makes its members feel safe enough from wild animals, extremes of temperature, criminals, assault and murder, tyranny, etc. Therefore, in a very real sense, the adult no longer has any safety needs as active motivators. Just as a sated man no longer feels hungry, a safe man no longer feels endangered. If we wish to see these needs directly and clearly, we must turn to neurotic or near-neurotic individuals, and to the economic and social underdogs. In between these extremes, we can perceive the expressions of safety needs only in such phenomena as, for instance, the common preference for a job with tenure and protection, the desire for a savings account, and for insurance of various kinds (medical, dental, unemployment, disability, old age).

Other broader aspects of the attempt to seek safety and stability in the world are seen in the very common preference for familiar rather than unfamiliar things, or for the known rather than the unknown. The tendency to have some religion or world-philosophy that organizes the universe and the people in it into some sort of satisfactorily coherent, meaningful whole is also in part motivated by safety-seeking. Here too we may list science and philosophy in general as partially motivated by the safety needs (we shall see later that there are also other motivations to scientific, philosophical, or religious endeavor).

Otherwise the need for safety is seen as an active and dominant mobilizer of the organism's resources only in emergencies, e.g., war, disease, natural catastrophes, crime waves, societal disorganization, neurosis, brain injury, chronically bad situation.

Some neurotic adults in our society are, in many ways, like the unsafe child in the desire for safety, although in adults it takes on a somewhat special appearance. Their reaction is often to unknown, psychological dangers in a world that is perceived to be hostile, overwhelming, and threatening. Such a person behaves as if a great catastrophe were almost always impending, i.e., usually responding as if to an emergency. Safety needs often find specific expression in a search for a protector, or a stronger person on whom to depend, or perhaps a Führer.

The neurotic individual may be described in a slightly different way with some usefulness as a grownup person who retains childish attitudes toward the world. That is to say, a neurotic adult may be said to behave "as if" actually afraid of a spanking, or of the mother's disapproval, or of being abandoned by the parents, or having food taken away. It is as if childish attitudes of fear and threat reaction to a dangerous world had gone underground and, untouched by the growing up and learning processes, were now ready to be called out by any stimulus that would make a child feel endangered and threatened. (Not all neurotic individuals feel unsafe. Neurosis may have at its core a thwarting of the affection and esteem needs in a person who is generally safe.)

The neurosis in which the search for safety takes its clearest form is in the compulsive-obsessive neurosis. Compulsive-obsessives try frantically to order and stabilize the world so that no unmanageable, unexpected, or unfamiliar dangers will ever appear. They hedge themselves about with all sorts of ceremonials, rules, and formulas so that every possible contingency may be provided for and so that no new contingencies may appear. They are much like the brain-injured cases who manage to maintain their equilibrium by avoiding everything unfamiliar and strange and by ordering their restricted world in such a neat, disciplined, orderly fashion that everything in the world can be counted upon. They try to arrange the world so that anything unexpected (dangers) cannot possibly occur. If, through no fault of their own, something unexpected does occur, they go into a panic reaction as if this unexpected occurrence constituted a grave danger. What we can see only as a none-too-strong preference in the healthy person, i.e., preference for the familiar, becomes a life-and-death necessity in abnormal cases.

THE LOVE NEEDS

If both the physiological and the safety needs are fairly well gratified, then there will emerge the love and affection and belongingness needs, and the whole cycle already described will repeat itself with this new center. Now the person will feel keenly, as never before, the absence of friends, or a sweetheart, or a wife, or children. He will hunger for affectionate relations with people in general, namely, for a place in the group, and will strive with great intensity to achieve this goal. He will want to attain such a place more than anything else in the world and may even forget that once, when he was hungry for food, he sneered at love.

In our society the thwarting of these needs is the most commonly found core in cases of maladjustment and more severe psychopathology. Love and affection, as well as their possible expression in sexuality, are generally looked upon with ambivalence and are customarily hedged about with many restrictions and inhibitions. Practically all theorists of psychopathology have stressed thwarting of the love needs as basic in the picture of maladjustment. Many clinical studies have therefore been made of this need and we know more about it perhaps than any of the other needs except the physiological ones.

One thing that must be stressed at this point is that love is not synonymous with sex. Sex may be studied as a purely physiological need. Ordinarily sexual behavior is multidetermined, that is to say, determined not only by sexual but also by other needs, chief among which are the love and affection needs. Also not to be overlooked is the fact that the love needs involve both giving *and* receiving love.

THE ESTEEM NEEDS

All people in our society (with a few pathological exceptions) have a need or desire for a stable, firmly based, (usually) high evaluation of themselves, for self-respect, or self-esteem, and for the esteem of others. By firmly based self-esteem, we mean that which is soundly based upon real capacity, achievement, and respect from others. These needs may be classified into two subsidiary sets. These are, first, the desire for strength, for achievement, for adequacy, for confidence in the face of the world, and for independence and freedom. Second, we have what we may call the desire for reputation or prestige (defining it as respect or esteem from other people), recognition, attention, importance, or appreciation. Perhaps the desire for prestige and respect from others is subsidiary to the desire for self-esteem or confidence in oneself. Observation of children seems to indicate that this is so, but clinical data give no clear support for such a conclusion. These needs have been relatively stressed by Adler and his followers, and have been relatively neglected by Freud and the psychoanalysts. More and more today, however, there is appearing widespread appreciation of their central importance.

Satisfaction of the self-esteem need leads to feelings of self-confidence, worth, strength, capability, and adequacy of being useful and necessary in the world. But the thwarting of these needs produces feelings of inferiority, of weakness, and of helplessness. These feelings in turn give rise to either basic discouragement or compensatory or neurotic trends. An appreciation of the necessity of basic self-confidence, and an understanding of how helpless people are without it, can be easily gained from a study of severe traumatic neurosis.

THE NEED FOR SELF-ACTUALIZATION

Even if all these needs are satisfied, we may still often (if not always) expect that a new discontent and restlessness will soon develop, unless the individual is doing what he or she is fitted for. A musician must make music, an artist must paint, a poet must write, to be ultimately happy. What a person *can* be, he or she *must* be. This need we may call self-actualization.

This term, first coined by Goldstein, is being used in this paper in a much more specific and limited fashion. It refers to the desire for self-fulfillment, namely, to the tendency for a person to become actualized in what he or she is potentially. This tendency might be phrased as the desire to become more and more what one is, to become everything that one is capable of becoming.

The specific form that these needs will take will of course vary greatly from person to person. In one individual it may take the form of the desire to be an ideal parent, in an another it may be expressed athletically, and in still another it may be expressed in painting pictures or in inventions. It is not necessarily a creative urge, although in people who have any capacities for creation it will take this form.

The clear emergence of these needs rests on prior satisfaction of the physiological, safety, love, and esteem needs. We shall call people who are satisfied in these needs, basically satisfied people, and it is from these that we may expect the fullest (and healthiest) creativeness. Since, in our society, basically satisfied people are the exception, we do not know much more about self-actualization, either experimentally or clinically. It remains a challenging problem for research.

THE PRECONDITIONS FOR THE BASIC NEED SATISFACTIONS

There are certain conditions that are immediate prerequisites for the basic need satisfactions. Danger to these is reacted to almost as if it were a direct danger to the basic needs themselves. Such conditions as freedom to speak, freedom to do what one wishes so long as no harm is done to others, freedom to express oneself, freedom to investigate and seek information, freedom to defend oneself, justice, fairness, honesty, orderliness in the group are examples of such preconditions for basic need satisfactions. Thwarting in these freedoms will be reacted to with a threat or emergency response. These conditions are not ends in themselves but they are *almost* so since they are so closely related to the basic needs, which are apparently the only ends in themselves. These conditions are defended because without them the basic satisfactions are quite impossible, or at least, very severely endangered.

If we remember that the cognitive capacities (perceptual, intellectual, learning) are a set of adjustive tools, which have among other functions, that of satisfaction of our basic needs, then it is clear that any danger to them, any deprivation or blocking of their free use, must also be indirectly threatening to the basic needs themselves. Such a statement is a partial solution of the general problems of curiosity, the search for knowledge, truth, and wisdom, and the ever-persistent urge to solve the cosmic mysteries.

We must therefore introduce another hypothesis and speak of degrees of closeness to the basic needs, for we have already pointed out that *any* conscious desires (partial goals) are more or less important as they are more or less close to the basic needs. The same statement may be made for various behavior acts. An act is psychologically important if it contributes directly to satisfaction of basic needs. The less directly it so contributes, or the weaker this contribution is, the less important this act must be conceived to be from the point of view of dynamic psychology. A similar statement may be made for the various defense or coping mechanisms. Some are very directly related to the protection or attainment of the basic needs, others are only weakly and distantly related. Indeed if we wished, we could speak of more basic and less basic defense mechanisms, and then affirm that danger to the more basic defenses is more threatening than danger to less basic defenses (always remembering that this is so only because of their relationship to the basic needs).

THE DESIRES TO KNOW AND TO UNDERSTAND

So far, we have mentioned the cognitive needs only in passing. Acquiring knowledge and systematizing the universe have been considered as, in part, techniques for the achievement of basic safety in the world, or, for the intelligent person, expressions of self-actualization. Also freedom of inquiry and expression have been discussed as preconditions of satisfactions of the basic needs. True though these formulations may be, they do not constitute definitive answers to the question as to the motivation role of curiosity, learning, philosophizing, experimenting, etc. They are, at best, no more than partial answers.

This question is especially difficult because we know so little about the facts. Curiosity, exploration, desire for the facts, desire to know may certainly be observed easily enough. The fact that they often are pursued even at great cost to the individual's safety is an earnest of the partial character of our previous discussion. In addition, I must admit that, though I have sufficient clinical evidence to postulate the desire to know as a very strong drive in intelligent people, no data are available for unintelligent people. It may then be largely a function of relatively high intelligence. Rather tentatively, then, and largely in the hope of stimulating discussion and research, we shall postulate a basic desire to know, to be aware of reality, to get the facts, to satisfy curiosity, or as Wertheimer phrases it, to see rather than to be blind.

This postulation, however, is not enough. Even after we know, we are impelled to know more and more minutely and microscopically on the one hand, and on the other, more and more extensively in the direction of a world-philosophy, religion, etc. The facts that we acquire, if they are isolated or atomistic, inevitably get theorized about, and either analyzed or organized or both. This process has been phrased by some as the search for "meaning." We shall then postulate a desire to understand, to systematize, or organize, to analyze, to look for relations and meanings.

Once these desires are accepted for discussion, we see that they too form themselves into a small hierarchy in which the desire to know is prepotent over the desire to understand. All the characteristics of a hierarchy of prepotency that we have described above seem to hold for this one as well.

We must guard ourselves against the too easy tendency to separate these desires from the basic needs we have discussed above, namely, to make a sharp dichotomy between "cognitive" and "conative" needs. The desire to know and to understand are themselves conative, i.e., have a striving character, and are as much personality needs as the "basic needs" we have already discussed.

FURTHER CHARACTERISTICS OF THE BASIC NEEDS

THE DEGREE OF FIXITY OF THE HIERARCHY OF BASIC NEEDS

We have spoken so far as if this hierarchy were a fixed order but actually it is not nearly as rigid as we may have implied. It is true that most of the people with whom we have worked have seemed to have these basic needs in about the order that has been indicated. However, there have been a number of exceptions.

1. There are some people in whom, for instance, self-esteem seems to be more important than love. This most common reversal in the hierarchy is usually due to the development of the notion that the person who is most likely to be loved is a strong or powerful person, one who inspires respect or fear, and who is self-confident or aggressive. Therefore such people who lack love, and seek it, may try hard to put on a front of aggressive, confident behavior. But essentially they seek high self-esteem and its behavior expressions more as a means to an end than for its own sake; they seek self-assertion for the sake of love rather than for self-esteem itself.

2. There are other, apparently innately creative people in whom the drive to creativeness seems to be more important than any other counterdeterminant. Their creativeness might appear not as self-actualization released by basic satisfaction, but in spite of lack of basic satisfaction.

3. In certain people the level of aspiration may be permanently deadened or lowered. That is to say, the less prepotent goals may simply be lost, and may

disappear forever, so that people who have experienced life at a very low level, e.g., chronic unemployment, may continue to be satisfied for the rest of their lives if only they can get enough food.

4. The so-called psychopathic personality is another example of permanent loss of the love needs. These are people who, according to the best data available, have been starved for love in the earliest months of their lives and have simply lost forever the desire and the ability to give and to receive affection (as animals lose sucking or pecking reflexes that are not exercised soon enough after birth).

5. Another cause of reversal of the hierarchy is that when a need has been satisfied for a long time, this need may be underevaluated. People who have never experienced chronic hunger are apt to underestimate its effects and to look upon food as a rather unimportant thing. If they are dominated by a higher need, this higher need will seem to be the most important of all. It then becomes possible, and indeed does actually happen, that they may, for the sake of this higher need, put themselves into the position of being deprived in a more basic need. We may expect that after a long-time deprivation of the more basic need there will be a tendency to reevaluate both needs so that the more prepotent need will actually become consciously prepotent for the individual who may have given it up very lightly. Thus, a man who has given up this job rather than lose his self-respect, and who then "starves" for six months or so, may be willing to take his job back even at the price of losing his self-respect.

6. Another partial explanation of *apparent* reversals is seen in the fact that we have been talking about the hierarchy of prepotency in terms of consciously felt wants or desires rather than of behavior. Looking at behavior itself may give us the wrong impression. What we have claimed is that the person will *want* the more basic of two needs when deprived in both. There is no necessary implication here that the person will act upon such desires. Let us say again that there are many determinants of behavior other than the needs and desires.

7. Perhaps more important than all these exceptions are the ones that involve ideals, high social standards, high values, and the like. With such values people become martyrs; they will give up everything for the sake of a particular ideal, or value. These people may be understood, at least in part, by reference to one basic concept (or hypothesis) which may be called "increased frustration-tolerance through early gratification." People who have been satisfied in their basic needs throughout their lives, particularly in their earlier years, seem to develop exceptional power to withstand present or future thwarting of these needs simply because they have

strong, healthy character structure as a result of basic satisfaction. They are the "strong" people who can easily weather disagreement or opposition, who can swim against the stream of public opinion, and who can stand up for the truth at great personal cost. It is just the ones who have loved and been well loved, and who have had many deep friendships, who can hold out against hatred, rejection, or persecution.

I say all this despite the fact that there is a certain amount of sheer habituation that is also involved in any full discussion of frustration tolerance. For instance, it is likely that those persons who have been accustomed to relative starvation for a long time are partially enabled thereby to understand food deprivation. What sort of balance must be made between these two tendencies, of habituation on the one hand, and of past satisfaction breeding present frustration tolerance on the other hand, remains to be worked out by further research. Meanwhile we may assume that they are both operative, side by side, since they do not contradict each other. In respect to this phenomenon of increased frustration-tolerance, it seems probable that the most important gratifications come in the first two years of life. That is to say, people who have been made secure and strong in the earliest years tend to remain secure and strong thereafter in the face of whatever threatens.

DEGREES OF RELATIVE SATISFACTION

So far, our theoretical discussion may have given the impression that these five sets of needs are somehow in a stepwise, all-or-none relationship to each other. We have spoken in such terms as the following: "If one need is satisfied, then another emerges." This statement might give the false impression that a need must be satisfied 100 percent before the next need emerges. In actual fact, most members of our society who are normal are partially satisfied in all their basic needs and partially unsatisfied in all their basic needs at the same time. A more realistic description of the hierarchy would be in terms of decreasing percentages of satisfaction as we go up the hierarchy of prepotency. For instance, if I may assign arbitrary figures for the sake of illustration, it is as if the average citizen is satisfied perhaps 85 percent in physiological needs, 70 percent in safety needs, 50 percent in love needs, 40 percent in self-esteem needs, and 10 percent in self-actualization needs.

As for the concept of emergence of a new need after satisfaction of the prepotent need, this emergence is not a sudden saltatory phenomenon but rather a gradual emergence by slow degrees from nothingness. For instance, if prepotent need A is satisfied only 10 percent, then need B may not be visible at all. However, as this need A becomes satisfied 25 percent, need B may emerge 5 percent; as need A becomes satisfied 75 percent, need B may emerge 90 percent, and so on.

UNCONSCIOUS CHARACTER OF NEEDS

These needs are neither necessarily conscious nor unconscious. On the whole, however, in the average person, they are more often unconscious rather than conscious. It is not necessary at this point to overhaul the tremendous mass of evidence that indicates the crucial importance of unconscious motivation. It would by now be expected, on a priori grounds alone, that unconscious motivations would on the whole be rather more important than the conscious motivations. What we have called the basic needs are very often largely unconscious although they may, with suitable techniques and sophisticated people, become conscious.

CULTURAL SPECIFICITY AND GENERALITY OF NEEDS

The classification of basic needs makes some attempt to take account of the relative unity behind the superficial differences in specific desires from one culture to another. Certainly in any particular culture an individual's conscious motivational content will usually be extremely different from the conscious motivational content of an individual in another society. However, it is the common experience of anthropologists that people, even in different societies, are much more alike than we would think from our first contact with them, and that as we know them better we seem to find more and more of this commonness. We then recognize the most startling differences to be superficial rather than basic, e.g., differences in style of hairdress, clothes, tastes in food, etc. Our classification of basic needs is in part an attempt to account for this unity behind the apparent diversity from culture to culture. No claim is made that it is ultimate or universal for all cultures. The claim is made only that it is relatively *more* ultimate, more universal, more basic, than the superficial conscious desires from culture to culture, and makes a somewhat closer approach to common human characteristics. Basic needs are *more* common-human than superficial desires or behaviors.

MULTIPLE MOTIVATIONS OF BEHAVIOR

These needs must be understood *not* to be *exclusive* or single determiners of certain kinds of behavior. An example may be found in any behavior that seems to be physiologically motivated, such as eating or sexual play or the like. The clinical psychologists have long since found that any behavior may be a channel through which flow various determinants. Or to say it in another way, most behavior is multi-motivated. Within the sphere of motivational determinants any behavior tends to be determined by several or *all* of the basic needs simultaneously rather than by only one of them. The latter would be more an exception than the former. Eating may be partially for the sake of filling the stomach, and partially for the sake of comfort and amelioration of other needs. One may make love not only for pure sexual release, but also to convince oneself of one's masculinity or femininity, or to make a conquest, to feel powerful, or to win more basic affection. As an illustration, I may point out that it would be possible (theoretically if not practically) to analyze a single act of an individual and see in it the expression of physiological needs, safety needs, love needs, esteem needs, and self-actualization. This contrasts sharply with the more naive brand of trait psychology in which one trait or one motive accounts for a certain kind of act, e.g., an aggressive act is traced solely to a trait of aggressiveness.

MULTIPLE DETERMINANTS OF BEHAVIOR

Not all behavior is determined by the basic needs. We might even say that not all behavior is motivated. There are many determinants of behavior other than motives. For instance, one other important class of determinants is the so-called field determinants. Theoretically, at least, behavior may be determined completely by the field, or even by specific isolated external stimuli, as in association of ideas, or certain conditioned reflexes. If in response to the stimulus word "table," I immediately perceive a memory image of a table, this response certainly has nothing to do with my basic needs.

Second, we may call attention again to the concept of "degree of closeness to the basic needs" or "degree of motivation." Some behavior is highly motivated, other behavior is only weakly motivated. Some is not motivated at all (but all behavior is determined).

Another important point is that there is a basic difference between expressive behavior and coping behavior (functional striving, purposive goal seeking). An expressive behavior does not try to do anything; it is simply a reflection of the personality. A stupid man behaves stupidly, not because he wants to, or tries to, or is motivated to, but simply because he *is* what he is. The same is true when I speak in a bass voice rather than tenor or soprano. The random movements of a healthy child, the smile on the face of a happy man even when he is alone, the springiness of the healthy woman's walk, are other examples of expressive, nonfunctional behavior. Also the *style* in which a person carries out almost all his or her behavior, motivated as well as unmotivated, is often expressive.

We may ask, is *all* behavior expressive or reflective

of the character structure? The answer is no. Rote, habitual, automatized, or conventional behavior may or may not be expressive. The same is true for most "stimulus-bound" behaviors.

It is finally necessary to stress that expressiveness of behavior and goal-directedness of behavior are not mutually exclusive categories. Average behavior is usually both.

GOALS AS CENTERING PRINCIPLE IN MOTIVATION THEORY

It will be observed that the basic principle in our classification has been neither the instigation nor the motivated behavior but rather the functions, effects, purposes, or goals of the behavior. It has been proven sufficiently by various people that this is the most suitable point for centering in any motivation theory.

ANIMAL-CENTERING AND HUMAN-CENTERING

This theory starts with the human being rather than any lower and presumably "simpler" animal. Too many of the findings that have been made in animals have been proven to be true for animals but not for the human being. There is no reason whatsoever why we should start with animals in order to study human motivation. The logic or rather illogic behind this general fallacy of "pseudosimplicity" has been exposed often enough by philosophers and logicians as well as by scientists in each of the various fields. It is no more necessary to study animals before one can study humans than it is to study mathematics before one can study geology or psychology or biology.

We may also reject the old, naive behaviorism that assumed it was somehow necessary, or at least more "scientific," to judge human beings by animal standards. One consequence of this belief was that the whole notion of purpose and goal was excluded from motivational psychology simply because one could not ask a white rat about its purposes.

MOTIVATION AND THE THEORY OF PSYCHOPATHOGENESIS

The conscious motivational content of everyday life has, according to the foregoing, been conceived to be relatively important or unimportant accordingly as it is more or less closely related to the basic goals. A desire for an ice cream cone might actually be an indirect expression of a desire for love. If it is, then this desire for the ice cream cone becomes extremely important motivation. If, however, the ice cream is simply something to cool the mouth with, or a casual appetitive reaction, then the desire is relatively unimportant. Everyday conscious desires are to be regarded as symptoms, as surface indicators of more basic needs. If we were to take these superficial desires at their face value, we would find ourselves in a state of complete confusion that could never be resolved, since we would be dealing seriously with symptoms rather than with what lay behind the symptoms.

Thwarting of unimportant desires produces no psychopathological results; thwarting of a basically important need does produce such results. Any theory of psychopathogenesis must then be based on a sound theory of motivation. A conflict or a frustration is not necessarily pathogenic. It becomes so only when it threatens or thwarts the basic needs, or partial needs that are closely related to the basic needs.

THE ROLE OF GRATIFIED NEEDS

It has been pointed out above several times that our needs usually emerge only when more prepotent needs have been gratified. Thus gratification has an important role in motivation theory. Apart from this, however, needs cease to play an active determining or organizing role as soon as they are gratified.

What this means is that, e.g., a basically satisfied person no longer has the needs for esteem, love, safety, etc. The only sense in which he or she might be said to have them is in the almost metaphysical sense that a sated person has hunger, or a filled bottle has emptiness. If we are interested in what *actually* motivates us, and not in what has, will, or might motivate us, then a satisfied need is not a motivator. It must be considered for all practical purposes simply not to exist, to have disappeared. This point should be emphasized because it has been either overlooked or contradicted in every theory of motivation I know. The perfectly healthy, normal, fortunate human has no sex needs or hunger needs, or needs for safety, or for love, or for prestige, or self-esteem, except in stray moments of quickly passing threat. If we were to say otherwise, we should also have to aver that every human had all the pathological reflexes, because if the nervous system were damaged, these would appear.

It is such considerations as these that suggest the bold postulation that a person who is thwarted in any of the basic needs may fairly be envisaged simply as sick. This is a fair parallel to our designation of the person who lacks vitamins or minerals as "sick." Who is to say that a lack of love is less important than a lack of vitamins? Since we know the pathogenic effects of love starvation, who is to say that we are invoking value-questions in an unscientific or illegitimate way, any more than the physician does who diagnoses and treats pellegra or scurvy? If I were permitted this usage, I should then say simply that a healthy person is primarily motivated by the need to develop and actualize his or her fullest potentialities and capacities. If a person has any other basic needs in any active, chronic sense, then he or she is simply unhealthy, as surely sick as if suddenly developing a strong salt hunger or calcium hunger. If we were to use the word "sick" in this way, we should then also have to face squarely the relations of humans to their society. One clear implication of our definition would be that (1) since a man is to be called sick who is basically thwarted, and (2) since such basic thwarting is made possible ultimately only by forces outside the individual, then (3) sickness in the individual must come ultimately from a sickness in the society. The "good" or healthy society would then be defined as one that permitted man's highest purposes to emerge by satisfying all the prepotent basic needs.

If this statement seems unusual or paradoxical, we may be assured that this is only one among many such paradoxes that will appear as we revise our ways of looking at deeper human motivations. When we ask what a person wants of life, we deal with his or her very essence.

SUMMARY

1. There are at least five sets of goals, which we may call basic needs. These are, briefly, physiological, safety, love, esteem, and self-actualization. In addition, we are motivated by the desire to achieve or maintain the various conditions upon which these basic satisfactions rest and by certain more intellectual desires.

2. These basic goals are related to each other, being arranged in a hierarchy of prepotency. This means that the most prepotent goal will monopolize consciousness and will tend of itself to organize the recruitment of the various capacities of the organism. The less prepotent needs are minimized, even forgotten or denied. But when a need is fairly well satisfied, the next prepotent ("higher") need emerges, in turn to dominate the conscious life and to serve as the center of organization of behavior, since gratified needs are not active motivators.

Thus humans are perpetually wanting animals. Ordinarily the satisfaction of these wants is not altogether mutually exclusive, but only tends to be. The average member of our society is most often partially satisfied and partially unsatisfied in all of his or her wants. The hierarchy principle is usually empirically observed in terms of increasing percentages of nonsatisfaction as we go up the hierarchy. Reversals of the average order of the hierarchy are sometimes observed. Also, it has been observed that an individual may permanently lose the higher wants in the hierarchy under special conditions. There are not only ordinarily multiple motivations for usual behavior, but in addition many determinants other than motives.

3. Any thwarting or possibility of thwarting of these basic human goals, or danger to the defenses which protect them, or to the conditions upon which they rest, is considered to be a psychological threat. With a few exceptions, all psychopathology may be partially traced to such threats. A basically thwarted person might actually be defined as "sick," if we wish.

4. It is such basic threats that bring about the general emergency reactions.

Chapter 3. Study Guide Questions

1. What does Maslow mean by the statement that "the basic human needs are organized into a hierarchy of relative prepotency"?

2. After physiological needs are gratified, what are the next needs to emerge?

3. Why are safety needs usually not motivators in our society?

4. If both physiological and safety needs are gratified, what are the next needs to be felt?

5. A. What feelings does satisfaction of a person's self-esteem needs lead to?

B. What happens if a person's self-esteem needs are not satisfied?

6. A. What are the highest needs that emerge after satisfaction of physiological, safety, love, and esteem needs?

B. What form does satisfaction of these needs take?

7. What are the most common exceptions to the hierarchy of needs as postulated by Maslow?

8. Which people seem best able to withstand nongratification of various needs in the hierarchy?

9. Is it necessary that all lower needs be 100 percent satisfied in order for higher needs to become motivators?

10. What does it mean to say that most behavior is multimotivated?

11. What effect does a satisfied need have on motivation?

Vocabulary

physiological	stimulus
hierarchy	maladjustment
equilibrium	cognitive
contingency	

Chapter 4.   Leadership Is Flesh and Blood

By Adm. Carlisle A. H. Trost

THE CHALLENGE OF LEADERSHIP

Today, leaders at all levels in the Navy face a different set of challenges, more complicated and in some ways more difficult than our predecessors knew. The challenges are significant, but we can overcome them. Lessons from the past, principles learned from our own experience, and the advice and counsel of other leaders can help guide us. We must make time to bring these home to our people. The values that inspire them to do their best are eternal. In particular, we must concentrate on four intangibles.

ONE

Leadership must flow from strength, both physical and moral. This begins with the leader's personal example and does not end until the organization comprehends, accepts, complies with, and resonates with the highest possible standards. Physical strength means physical fitness and more. Under the

Source: U.S. Naval Institute *Proceedings* (February 1988): 78–81.

exceptional stresses to which we subject our people, leaders' vitality must be the last to fail. When the issue is in doubt, frequently a subordinate will look to the leader to see how things are going. If the leader is positive, subordinates will try harder with a greater will to win. Ultimately, the leader's own energy will return to him or her through them, and the leader will be stronger.

Moral strength cannot be quantified, but is easily felt within the command. It begins at the point where people accept that leaders must know what they are talking about. If the leader does know and is therefore not to be taken lightly, the next concern is whether he or she is empowered by the right motives. The leader must always work from an ethical base, for the art of leadership depends on value judgments. The leader's personal ability to discriminate between right and wrong may be the only resource available. Time may not permit consultation with others. The leader must stand up for personal beliefs, even if it means standing alone. The expedient and the right courses of action may coincide; if not, the leader must choose. And we must prepare leaders for their choices by reinforcing, throughout their career, the ethical base as the source of decisions.

TWO

Leadership must deal with reality, the reality of the sailors and the equipment in the spaces, not the monitoring systems inside the neat, air-conditioned, soundproofed consoles. All the technology in the world cannot substitute for getting out and observing things firsthand, for finding the inconvenient details that would otherwise merge into the background. This should be done both formally and informally, through programs like zone and personnel inspections and through the time-tested practice of being with one's people.

The pictures that emerge from such experiences may differ widely, for we cannot assume that reality will always fit into familiar patterns. Leaders must learn to deal with various levels of disorder. A Harpoon engagement may have neatly defined parameters; the discipline problem is unlikely to be so neat; and as we saw in 1987 in the aftermath of the attack against the USS *Stark* (FFG-31), the response to fire or flooding or other emergencies may require the book to be thrown out. But whether preplanned or improvised or something in between, all actions must begin with an understanding of what in fact is taking place, and all must be approached with a zeal to get to the heart of the matter and apply one's best energies against it.

THREE

Leadership must make the most of all available assets. As leaders, we must consider the individual in every aspect—not only as a provider of services but also as a living entity. Knowing your people means much more than simply memorizing a few facts about them. It means knowing their capabilities and limitations, their ambitions, how they communicate, and how they approach a problem. It means developing a sensitivity to all subordinates, including minorities, so that we can ensure that they are full participants within the command.

We must never forget that, with all of their strengths and weaknesses, our subordinates are under our charge 24 hours a day. Many of them are young, and perhaps on their first tour of duty. Many are married. Their parents, guardians, or spouses have lent them to us with a faith that the Navy "system" will look after their welfare and safety, help them through the maturing process, and make them better citizens through challenge and responsibility. That, too, is a 24-hour-a-day charge. We live up to it by showing our interest in our subordinates from the moment they first cross the quarterdeck. We take time to train them. We give them the resources to do the job. We instruct them if they fail and recognize them if they succeed. We go to court

with them if they get into scrapes. We visit them in the hospital. We concern ourselves with their medical and dental care, and—almost obsessively—with their safety, both on duty and off, and particularly on the highways. And we want them to succeed, and expect that they will.

More than anything else, we must decipher and explain the situation so that all our subordinates will understand our perspective and will have the best chance of acting on it. This is one of the most significant changes from leadership in the past. The Navy is not a democratic institution, but implicit in any task is the vital necessity of understanding why it is to be performed. When trained, highly motivated subordinates—with whom our service is blessed in great numbers today—understand the why of a task, if they confront a change in circumstances, they will be able to reconstruct the what and the how even if we aren't there to guide them. And they will be motivated to perform that task to the highest standards we set. The great strength of our people today lies in this: give them a challenge, and they will not stop until they master it.

FOUR

Good leadership must provide that spark of relevance that foretells success. Human beings respond to clear direction; they accomplish tasks under good management; but they will give their all—and even more than their all—to the leader who stirs their blood, shows them how unique and remarkable they are, and how valuable their contribution. The obligation of a leader is to be successful. But if the leader can find the right word to animate the moment as well, the task will be that much easier. Leadership depends on good judgment, cool and detached, but it expresses itself through living people, not caricatures, whose emotions become engaged in carrying out the Navy's vital missions.

MASTERING THE CHALLENGE

As we enter a new era of constrained resources, but with no constraint on explosive technological growth, many assume that the role of people has diminished. That is not my view, and I am determined that whatever the size and composition of the Navy in the future, the quality of our personnel, our support for them, and our leadership will not change. The challenges may be more complicated today, but the principles that we rely on are the same that saw our predecessors through many wars and intervening periods of peace, through recoveries of defense strength and relative periods of decline. Now and in the future, if technology is to serve humankind, people must come first.

Chapter 4. Study Guide Questions

1. According to Admiral Trost, what are the four intangibles upon which leadership should be based?
2. Where does moral strength begin?
3. What must a leader do to be sure of getting the real facts about a problem?
4. What is meant by "knowing one's people"?
5. According to Admiral Trost, what has significantly changed about leadership today as opposed to the past?

6. To what kind of leader will subordinates respond best and give their all?

Vocabulary

predecessor	expedient
intangibles	console
comprehension	entity
resonate	decipher
vitality	perspective
ethical code	animate
value system	caricature

Chapter 5. Look, Listen, and Lead

By Sgt. Maj. John L. Horton

The challenge of exercising effective leadership within the Marine Corps is of great importance. It demands our immediate—and collective—attention. No unit is impervious to this challenge.

Effective leadership uses the arts of communication and supervision to get the job done. Sounds simple enough, doesn't it? It should be understood, however, that effective leadership is an ongoing, growing, and learning process—not a finished product. But once competent leadership is firmly established, it is everlasting. For its continued growth and success, we need only provide it with a healthy environment and constant vigilance.

Communication, encompassing all kinds of interpersonal dynamics and human understanding, has never been more important than in today's sophisticated, technocratic Marine Corps. For a task to be efficiently accomplished, it must be communicated—through both written and spoken language—to the proper persons. Most of us understand this simple fact of life. Why, then, does it keep haunting us? An old proverb answers this question: "A thing must be heard and understood for it to be done." Enough said. As one advances up the leadership ladder, communication skills become more useful than technical skills and expertise. That is why it is so important to become a competent communicator if one hopes to become an effective leader.

Supervision, the twin sister of communication, is a 24-hour, seven-day-per-week job. Someone has to be on duty and be fully accountable for his or her watch. Although it is a tough and sometimes thankless job, the effective leader has to get the job done with people. Supervision is thus a good barometer of a leader's competence: if the leader is in charge, it tells; if the leader is not in charge, it also tells. Either

Source: U.S. Naval Institute *Proceedings* (November 1987): 84–87.

the job gets done or it doesn't. Too often, compliance has to be monitored or the particular job is apt to be gaffed off. Involvement by appropriate authority usually means that the job gets done right the first time around. This is the essence of good supervision.

Proper supervision is also an instrument with which to positively reinforce effective leadership. It allows a good leader to catch someone doing something right, as well as to catch someone doing something wrong.

Moving beyond the basics, my definition of effective leadership would include, but not be limited to, competence or expertise in a specific field or activity, adherence to high standards of excellence, and an ability to get the job done through people—by getting others to do what you want them to do, for a common cause.

Few will argue that the Marine Corps needs more and better leaders. But where do we find them or how do we create them? The obvious answer is that we must create environments that are permeated with effective leadership. We already have many of these environments: recruit training, officer candidate school, the drill field, noncommissioned officer's (NCO) school, staff noncommissioned officer's (SNCO) academy, certain formal and military occupational specialty schools, and certain other duty assignments. But these select few are the exceptions rather than the norm. Nevertheless, they provide correct role models for the rest—the great majority—of us.

I firmly believe there are no bad followers—just bad leaders. Generally speaking, Marines will go down the path that they are led. Therefore, it is up to the leaders to choose the right path. Capable leaders are endowed with certain definable traits and qualities. They are not yet perfect, because they are still growing and learning about themselves and others. But they personify the effective leader in technical skills as well as in human understanding. They exercise their strengths and improve upon their weaknesses. They are combination manager-leaders: managing things and leading people. They set the example. They take care of their people.

To effective leaders, the titles of Marine, NCO, SNCO, and leader mean something special. Such leaders are fully aware that effective leadership in today's Corps calls for sacrifice and service on the part of the leader on behalf of his or her Marines. Such leaders give a full-time commitment. They have know-how and a can-do attitude. Effective leaders understand that the attitudes and actions of those they lead are a direct reflection of their own. They understand, furthermore, that the buck does, truly, stop with them. They know that they owe it to themselves and their Marines to give—and to demand—the very best.

While daily work routines and mission priorities are realities for all of us, let us not forget that everything and everybody works more smoothly and efficiently in an environment where there is effective leadership. You must believe this, for everything else depends on this simple fact. And creating the right environment is up to all of us. If we have inefficient leaders, then we are the solution. Quite simply, if we do not become a part of the answer, we are a part of the problem.

The ingredients of a competitive and competent environment for our Marines are sharing and teamwork. Among other things, this means identifying and rewarding competence and excellence. This also means identifying substandard performers who need correction, closer supervision, and—if necessary—separation from the Corps. It should always be remembered that healthy competition and demonstrated competence breed excellence in all of us. Sharing and teamwork mean giving of oneself—one's talents, skills, expertise, knowledge—to make the team better. This means sharing the many professional secrets, formulas, and techniques with others. It also means helping others to develop into better professionals. Sharing and teamwork are good and right and honorable, and strengthen our beloved Corps.

Let me share some of the secrets and formulas that have worked for me over the years. I believe that effective leaders follow several basic rules.

Know Your Marines. Get to know them on both a personal and a professional basis. Try to learn what makes them tick. It is a never-ending job, an ongoing process that requires much time, practice, and patience. This is another example of one of those continuous, omnipresent responsibilities—not just an 0730–1630 job. It never has been easy; it never will be easy. It is, however, the foundation upon which leadership is built.

Talk to Your Marines. Get them together on a regular basis—all of them, no exceptions—and talk with and listen to them, sincerely and seriously. You would be surprised at what you can learn about each other and about the organization as a whole. Many of the Corps's most difficult and pressing problems have been solved or resolved by listening to the troops. This kind of communication can take place in a formal setting (monthly leadership meetings, breakfasts, luncheons, and the like) or in an informal setting (wandering around, casual conversation, impromptu meetings, and the like).

Establish and Maintain Standards of Excellence. Tell your Marines what you expect of them, and make it perfectly clear to everyone. Set clear procedures and standards for attaining short-range objectives and long-range goals. Everyone should be expected to—and allowed to—perform his or her job. Where necessary and practical, specific guidance and individual counseling should be introduced to enable a Marine to become a better Marine. This is the essence of a helping relationship—helping a person to help himself or herself. Establishing and maintaining standards of excellence are at the heart of organizational development and effective leadership. These standards must be implemented strongly and vividly, otherwise exhibitions of neutrality, mediocrity, and/or negativity will become unpleasant realities. In terms of expectations, only three varieties are available: no expectations, low expectations, or high expectations. I prefer the last of the three as a working standard of excellence.

Watch for Telltale Signs. Learn to recognize both the positive and the negative signals of your Marines. Watch for physical, verbal, and nonverbal signs. They all tell you something. There is no neutral ground here. Either your Marines are gaining or they are losing ground; they are either accomplishing or not accomplishing their mission. There is no such thing as breaking even in the real world of effective leadership, and believing that there is will eventually get you into serious trouble. Also, learn to distinguish between human and technical problems. There is a difference. Again, I believe that human problems are more debilitating and vexing than most technical ones. While people are our most valuable resource, they are, unquestionably, the most complex of our assets to manage and nurture.

Enhance the "Win" Capability for the Individual and the Organization. Appreciate the fact that most people want to be winners. Also, appreciate the fact that it takes lots of camaraderie, competence, and creativity to get the job done. Mission accomplishment and personnel productivity do not come easy. They never have and never will. Good leaders owe it to their Marines to put them in situations where they will win. Furthermore, it is of paramount importance that everyone contributes to the mission. Contributing makes people feel good about themselves, about others, and about the organization as a whole. By teaching your Marines to win, you will help them gain more confidence and achieve greater success. By creating an atmosphere of team spirit and teamwork, you can mold your Marines into a true band of brothers and sisters.

I hope that I leave no false sense of a quick solution to this complex problem of creating the right leadership environment. Words alone are empty, but one has to begin somewhere—with a plan of action. How about it? Remember, there are basically three types of people: talkers, watchers, and doers. I ask: which type are you?

Chapter 5. Study Guide Questions

1. According to Sergeant Horton, which arts are used in practicing effective leadership?

2. Why is it important to be able to communicate well?

3. What is the essence of good supervision?

4. What are three characteristics of effective leadership?

5. What are the traits of an effective leader, according to Sergeant Horton?

6. What is involved in sharing and teamwork?

7. What are the basic rules that effective leaders follow?

8. What are the beneficial effects obtained when a person makes a real contribution to the unit mission?

Vocabulary

impervious	debilitating
interpersonal	camaraderie
dynamics	creativity
technocratic	productivity

Chapter 6.   Positive Leadership Techniques

By Department of Leadership and Law, U.S. Naval Academy

A great number of leadership techniques, all based on well-established principles, can be employed in motivating personnel. These have been thoroughly tried out through generations of Navy and Marine Corps leadership and have proven to be highly effective. The leader's problem frequently is not so much knowing what leadership techniques are available as it is knowing which of them to employ in a specific situation. Only a careful analysis of the techniques and their underlying principles will eliminate this uncertainty.

LEADERSHIP TECHNIQUES FOR LEADERSHIP RESPONSIBILITIES

Every act of leadership should make followers feel that as long as they are doing their best to follow, they will be secure and their efforts will receive recognition. There are a great many techniques that the leader can use to encourage these feelings, and an attempt will be made here to list these techniques according to the nature of the leadership act. For instance, a leader who forms the habit of giving

Source: Karel Montor and Anthony J. Ciotti, eds. *Fundamentals of Naval Leadership*. 3d ed. Annapolis: Naval Institute Press, 1984.

indefinite commands will soon discover that the resulting confusion in the minds of subordinates will make them lose confidence in that leader. Thus one technique in the giving of commands is that a command must be definite.

Though no all-inclusive list can be made, the techniques of leadership can be grouped into the following categories:

1. Giving commands
2. Giving orders
3. Getting cooperation
4. Establishing discipline
5. Improving morale that is low because of feelings of insecurity
6. Improving morale that is low because of feelings of lack of recognition
7. Properly using organization and administration.

The lists that follow should never be considered to be complete or established, and students may have occasion to add techniques that they find through experience to be workable and to fit to individual personalities. Additional techniques may become evident when various leadership situations arise, and they will be learned by repeated use.

LEADERSHIP TECHNIQUES INVOLVED IN GIVING COMMANDS

1. A command must be definite.

2. A command must be positive. It must be given in a tone of voice that leaves no doubt that it is to be executed.

3. The leader must look at subordinates when giving them a command.

4. A command must be concise. It must not be so long or involved that it cannot be remembered.

Unquestioning obedience to a command is the basic concept of military life. A command given by a proper authority demands the follower's immediate response to do the will of the leader. The objective of the disciplined way of life of any military school, such as the Naval Academy, is to indoctrinate the student with this concept. When a command is issued, there can be no question on the part of the follower as to whether the command is correct or whether there is any option other than carrying it out. Subordinates must be so trained that they immediately carry out the instructions contained in the command to the very best of their ability.

A command does not permit any question or discussion. Some of the familiar commands used in the Naval Service are "Forward, March!" "Right Full Rudder!" "All Engines Ahead Full!" and "Commence Firing!" Immediate and instinctive obedience is the only reaction possible to a properly worded and properly delivered command.

Commands become familiar as the result of constant usage. However, a command is not necessarily stereotyped. Situations that require commands often develop quickly and perhaps without precedent. For example, a junior officer inspecting the paint locker might discover a fire. In this situation the officer would immediately issue commands to personnel in the vicinity. One command, to the nearest enlisted person, might be "Report to the officer of the deck that there is a fire in the paint locker. On the double!" Another command, to a second enlisted person, might be "Get the fire extinguisher from the next compartment." In a situation such as this, command may be expected to follow command until the fire is extinguished. Every person receiving one of these commands has no alternative but to do exactly as told, immediately and without question. The combination of the leadership ability of commanders and the disciplined obedience of followers produces the team that wins competition in peacetime and victory in battle.

LEADERSHIP TECHNIQUES INVOLVED IN
GIVING ORDERS

1. Explain what is to be done. Discourage the tendency of the junior to ask how to do it, but leave an opening for questions of confused subordinates.

2. Don't talk down to the enlisted in giving instructions.

3. Give orders to the person in charge, and not to the group. The chain of command must be followed.

4. Encourage and coach the enlisted when they encounter difficulties.

5. Remember that the personnel are serving their country, not the officer as an individual.

6. In giving an order, try to get across the feeling of "Let's go!" instead of "Get going!"

7. Avoid an overbearing attitude.

8. Show confidence in the ability of subordinates.

9. Don't use a senior's name or rank to lend weight to your own order.

10. Give a reason for your order if time permits, or if it appears that the order will be clearer if subordinates understand the reason behind it.

LEADERSHIP TECHNIQUES INVOLVED IN
GETTING COOPERATION

1. Stimulate unit or organizational pride by showing your own pride and enthusiasm for the service.

2. Don't criticize another officer or another organization in the presence of subordinates.

3. Keep your subordinates informed so that they may have an intelligent sense of participation.

4. Use the word *we* instead of the word *I* whenever appropriate and possible.

5. Accept responsibility for corrections from higher authority, and take remedial action.

6. Give full credit to members of the organization whose work and ideas have brought progress.

7. Let your enlisted know that you think they are good, and maintain high standards through alert supervision.

8. Make sure that all subordinates know your policy.

9. Don't be sarcastic.

10. Don't threaten punishment to make an order effective.

11. Don't invent jobs just to keep subordinates busy.

LEADERSHIP TECHNIQUES INVOLVED IN
ESTABLISHING DISCIPLINE

1. Praise in public; censure in private.

2. Give subordinates the benefit of the doubt.

3. Punish the individual concerned, not the group.

4. Take into account whether or not an infraction of rules or regulations was intentional.

5. Consider a person's record.

6. Be impartial, consistent, and humane in giving rewards and punishment.

7. Never use severe punishment for minor offenses.

8. As soon as possible, remove senior subordinates who have demonstrated their unfitness.

9. Teach understanding of discipline rather than fear of it: punish the guilty promptly, and defend the innocent stoutly.

10. Support the correct actions of subordinates.

LEADERSHIP TECHNIQUES INVOLVED IN
IMPROVING FEELINGS OF SECURITY

1. Let subordinates know what is expected of them.

2. If you are pleased with their work, tell them so.

3. If possible, keep subordinates informed of what is in store for them.

4. Don't make promises you cannot keep.

5. Grant deserved favors willingly.

6. Know the state of the morale of your personnel.

7. Never "pick on" an individual!

8. Be certain that a subordinate understands why he or she is being disciplined.

9. Evaluate your own performance in terms of the individual morale and group esprit de corps that exists in your organization.

LEADERSHIP TECHNIQUES INVOLVED IN
GIVING RECOGNITION

1. Praise when praise is due. Don't flatter.

2. Be on the job whenever your subordinates are working.

3. Be interested in the promotion of your personnel. Encourage them to prepare for advancement.

4. See to it that you are the first person to whom a subordinate turns in case of trouble.

5. Express interest in ideas even though you might disagree with them.

6. Take a keen interest in the quarters and mess. Insist that these be the best available.

7. Study your personnel. Learn all about them: where they come from, their problems and interests, etc.

LEADERSHIP TECHNIQUES INVOLVED IN
IMPROVING ORGANIZATION AND ADMINISTRATION

1. Require use of the chain of command.

2. Conform to the rules of the organization.

3. Discover weaknesses of the organization by observing and questioning.

4. Never issue an order that is not going to be enforced.

5. Be fair about promotions.

6. Demote incompetents.

Chapter 6. Study Guide Questions

1. What is a major problem often faced by a leader?

2. According to this chapter, what are the seven categories into which the techniques of leadership can be grouped?

3. What are the four leadership techniques involved in giving commands?

4. What are the ten leadership techniques involved in giving orders?

5. What are the eleven leadership techniques involved in getting cooperation?

6. What are the ten leadership techniques involved in establishing discipline?

7. What are the nine leadership techniques involved in improving feelings of security?

8. What are the seven leadership techniques involved in giving recognition?

9. What are the six leadership techniques involved in improving organization and administration?

Vocabulary

generations	sarcastic
concise	censure
indoctrinate	infraction
stereotype	humane
overbearing	esprit de corps

The Responsibilities and Qualities of Leadership

Leaders

- know where they (and their group) are going and how to get there, or if they don't know how, they are willing to let their subordinates in on (1) the ultimate goal, (2) the proximate goals, and (3) the path(s) thereto, and to enlist their support.
- trust followers to be loyal, to have intelligence, to know their jobs and be able to carry them out without nitpicking supervision. Even if the subordinates do not possess all these characteristics, leaders treat them as if they do—and, surprisingly enough, they either will have them or will acquire them.
- treat followers like human beings, respecting feelings and abilities as well as limitations, treating them not as objects but as cooperative persons who are able to make useful contributions to the task(s) at hand.
- are truthful with followers. If leaders cannot explain all the reasons for commands they must give, they give as much information as possible and are not too proud to let subordinates know

that there may be factors beyond a leader's control.

- are in control of themselves, and actions, feelings, and emotions are not allowed to become excuses for venting frustration on others.
- do not expect that they have, or will have, all these qualities—but they're in there trying, even if this means letting subordinates in on the effort, letting them know their help is wanted and appreciated and will be accepted.
- practice—and see that their entourage does likewise—the elementary aspects of courtesy, such as saying "Thank you" and giving praise for jobs well done and for an individual's best efforts.

That is what leaders are. All others are managers. No one has ever followed a manager to the absolute limits, to the loss of life. And a leader, especially a military leader, must be able to ask this of followers. If a leader is what he or she is supposed to be, this will not be a problem. If that leader is not, no rules, regulations, structures, or coercive practices can force subordinates to pass this ultimate test of loyalty.

Chapter 1.   Leading by Example

There is a phrase describing the essence of a naval officer which has fallen into disuse. That phrase is "service reputation," the unwritten, unspoken, unlisted net assessment of an officer's pluses and minuses. When naval officers were less numerous than they are today, an officer knew most of the other officers and privately ranked them according to their service reputation. Perhaps the size and complexity of today's Navy and Marine Corps pre-

vent a "familiar" appraisal of and by each member of the officer corps. Perhaps modern management information systems distract officers from what truly matters to them about other officers, whether they be junior, senior, or peer. But, in their hearts, all officers know that their behavior and attitude have a profound effect on everyone they work with. Adm. Kazuomi Uchida agrees that

an officer's behavior considerably influences other individuals, particularly his subordinates. Regardless of an officer's personality or style, he must be always unselfish and fair to subordinates so that they never lose their reliance on him. An

Source: Adapted from chap. 2 of Karel Montor et al., eds. *Naval Leadership: Voices of Experience*. Annapolis: Naval Institute Press, 1987.

officer's humanity is perceived through his behavior, and it is this which moves others.

THE POWER OF POSITIVE RELATIONSHIPS

The relationship between officers and their people has an effect on everyone's performance. When crew members of a ship that has been deployed too long complain proudly about the hardships they are undergoing and develop a kind of spartan pride, they probably have a good relationship with their commander. The morale of a less well led crew can be expected to deteriorate progressively under exactly the same circumstances.

An officer can develop a great deal of camaraderie with subordinates without destroying the discipline that is so essential in any kind of an evolution, particularly in wartime. Adm. Elmo Zumwalt recounted this episode of his career:

I had the honor of commanding the world's first guided missile frigate, the USS *Dewey*, and because it was the first of the class, I was given the cream of the crop from the Bureau of Naval Personnel. The heads of departments were superb, and because they were superb and very sure of themselves, they were not the least bit hesitant to come up and discuss with me in very vigorous terms things that they thought could be done differently. But when the chips were down in any evolution and I gave the directive, there was instant and total obedience. The fun was always recognized as just part of the business of having a good team spirit.

In order to do an optimal job, an officer needs to believe that the Navy is worthwhile, that the service is something of which he is proud. He need not view the institution as something sacred and unchangeable, nor should he assume that every cog in that vast institution works with anything like perfection. He can be positively critical of those areas that need fixing, but he must have a dedication to the overall mission of the Navy and to the concept of the life of service in readiness to defend his country and be a good officer.

In the wardroom or in the staff, an officer should do his best to be a good corporate citizen of that unit. Where there are personality clashes, he should do his best to understand what makes the other fellow tick. There are compromises that need to be made in order for an officer to be a good overall team member, but there is also a very clear line beyond which he should retire when he runs into the occasional "bad apple." There may be someone on the ship or in a unit who has a very negative attitude and who can be extremely disruptive, and counseling and correc-

tive action regarding this individual is the responsibility of the officer next higher in that individual's chain of command.

THE POWER OF A POSITIVE ATTITUDE

In discussing the importance of a positive attitude, Adm. Thomas Hayward related this story:

There's a great story about Admiral Red Ramage and how he won his Medal of Honor with the *Parche*. He wouldn't stop and think about, "Gee, what will happen if we get hit or get sunk?" He kept in mind what he had to do, and that was to sink the enemy. It may have been in his mind sometimes in his tactical decisions about what is the risk he's taking, but if caution is the character of a person, that person's probably going to be too cautious for a fast-moving Navy tactical engagement. The way you overcome that, if you are an individual with a cautious nature (because you can be born that way), is to force yourself into training situations, and the Navy tries to do that, tries to expose you to enough training encounters that you develop self-confidence. Through meeting somebody, you want to help them arrive at that point, so in the debriefings you don't dwell on the person's negatives, except in the constructive way, and you help them accentuate what they did right, so they will get back out and do it again and keep working on it.

Now a few people wash out of flight training, you know, it's just going to happen. Some of them wash out because there's a physical inability to perform that particular kind of a mission, others wash out because they don't have the motivation; there's a lot of difference between whether you succeeded or didn't succeed in that sense. So it seems to me as though a TACCO (Tactical Coordinator in a P-3) has one and only one responsibility, and that's to learn his job so well he's got the confidence that he's going to hang in there, and if he loses a few, okay, but he makes the most of them and keeps a good reputation for pressing on.

An individual cannot be effective in any way or successful in any way at any level in the Navy if he does not have the capacity to take on the task of setting the example, and behavior is a basic element of setting the example. The crew has a right to look upon their leader with respect, but the leader has to earn their respect and work to keep it. The quickest way to lose respect is to set a bad example through gross behavior. This includes doing the kinds of things that some may think are macho or funny, but that in the long run are going to undermine the morale of the command and the leader's pride.

An officer's behavior, both on and off duty, is noticed by shipmates and helps define their perception of him or her as an officer and a leader. The famous naval officer Adm. Marc Mitscher is a prime example of a man who led by example. So great was Mitscher's concern for the training and welfare of his men that he was able to retain their final ounce of effort and loyalty. He was "a bulldog of a fighter, a strategist, and a seeker of truth." Similarly, officers cannot help but note the behavior of their people on and off the job, whether they are full-time personnel or drilling reservists. The old adage "an officer is on duty 24 hours a day" has even more significance in the era of the total force.

Working with and through others to further the mission of the command and, consequently, the mission of the Navy and Marine Corps is a vital aspect of an officer's job. Self-serving actions must be secondary. For example, although cooperating with other individuals and other organizations may be desirable to accomplish specific tasks, cooperation should not be an end in itself. That is, peer pressure to "go along" or "get along" with a situation that is ethically or morally wrong must be resisted. To yield to such pressure is to do a disservice to the Naval Service and to the service reputation of the officers involved. Admiral Hayward cautions that

> whatever his position of leadership, an individual who believes that he can conduct himself in a way that is not gentlemanly is making a significant error of judgment. Senior officers whose personal behavior was not "up to snuff" have never progressed very far in the Naval Service.

> A leader needs to work at getting along with people without compromising his beliefs. He can stand firm on an issue he feels strongly about without creating a problem, however. The boisterous, loud, tough actor is not the "right kind of guy" to follow. Generally speaking, while there are some who become very good officers who chew people up and spit them out with ease, the finest officers learn how to deal with every human being as a human being and can dress a person down properly, up in the right circumstances, and congratulate a person correctly under other circumstances. In short, the successful officer learns how to deal with difficult situations without ever compromising his principles in the process.

Recognizing the difficulty of reconciling the principles and ideals of the Naval Service with the reality of car payments and sick children is part of being a leader. A successful leader understands the personal history of each subordinate, knows from what segment of society each person is drawn, and understands the job of each person. A naval officer cannot begin to earn respect unless he or she possesses that kind of knowledge, and cannot expect to keep it unless his or her behavior is impeccable.

To Admiral Zumwalt, the episode that might be most useful in demonstrating the painful choices that an individual has to make in complex situations where duty is not necessarily clear transpired during the 1973 Arab–Israeli Yom Kippur war, when

> it became clear very early on that the Israelis were likely to be overrun because they were not able to count on the replacement of their equipment by us—we weren't moving the equipment forward fast enough. I went to see the Secretary of Defense [James Schlesinger], and told him that my judgment was that this was likely to happen. He said, "I agree, but my hands are tied." This seemed to me to be playing at war without being aware of how difficult it is to change the momentum of a battle. The President [Richard Nixon] had become quite unavailable to the chiefs by that time, and I therefore concluded that something had to be done and that I'd better do it.

> I went to see Senator [Henry] Jackson and gave him this information and urged him to call the President, which he did. The President directed that the equipment be immediately sent forward, which in my judgment made it possible to turn that war around. In what I did, I took action that the Secretary of Defense had not taken, and yet my conscience at the time led me to believe that it was the necessary action, and even in retrospect I believe that it was an appropriate thing to do, even though in so doing I certainly ran the risk that the judgment would be made that I would be dismissed from office. [Editor's note: It should be understood by the reader that this represents a decision made by an officer with decades of experience.]

Admiral Zumwalt tells of his experience as commander of our naval forces in Vietnam, when he presided over the command of the brown-water navy, coastal surveillance forces and the craft that were involved, and the rivers and canals and the aircraft assigned thereto.

> As one would expect on board ship, I found that in order really to be able to understand the situations that my very young brand-new naval officers, commanding officers, were experiencing in their small craft, and to be able to understand

how effectively to use them, I had to get out and be with them. So I spent a number of evenings in ambushes with them patrolling the rivers and canals. In discussions with them when we were out in ambush, I came to understand the limitations both of the equipment and of the people, and the very great capabilities of both, and I was able to visualize a lot more of the possibly otherwise unforeseeable contingencies with which we might have to deal in major campaigns operating out of Saigon.

Based on the knowledge that one gains from those operational inputs, it was possible to have detailed discussions with the personnel on my staff in Saigon and to prepare ourselves mentally for the fast decisions that we would have to make. Fast-breaking events took place when we took a new blockade along the Cambodian border and other places. I think that it's fair to say that we were able to foresee the general nature of almost every action or counteraction that the enemy might take. Incidentally, one of the things that we proved to ourselves was the age-old principle that you can get away with almost anything by surprise, and you can probably do the same thing a second time, but you better not try it the third, because by that time a wise enemy has pretty well learned to adapt to it.

The key to being a good staff officer is to have had the operational experience that makes it possible to understand the problems of the commander's subordinate operating units. The officer who goes to the staff of the admiral of a cruiser-destroyer flotilla is far more competent to help that admiral carry out his responsibilities if he has served on one of those cruisers or destroyers or a similar type. This is also true for those who are more logistics oriented, such as supply officers. A staff officer must also make frequent visits to the operating units and have a meticulous sense for those details that are essential to operational efficiency. He must be able to distinguish them from those details that could be categorized as simply minuscule.

A requirement for anyone who aspires to high rank is to read beyond the literature required for him to do well in his own particular job.

INSTILLING A POSITIVE ATTITUDE IN SUBORDINATES

After the Vietnam War, enlistment and reenlistment rates were way down. Admiral Zumwalt, who had been asked to address the problem, decided to approach it from the standpoint of morale.

I concluded that things were so bad, and the social fabric of the country that was supporting our armed services was so bad, that one had to take a revolutionary approach. I had for many years been waiting for my opportunity to strike a major blow in behalf of equal opportunity, and I was also a firm believer in the fact that women have as much capability as their male counterparts. I had just come from 20 months in command of our forces in Vietnam, where I had found that the most ruthless and cunning enemy I had ever had to face were the Vietcong women, and I concluded that dealing with the sexual differences in the Navy was timely.

It was also clear that to me, again as a result of my wartime exposure to the wonderful young men who were fighting in Vietnam, that there were some other very serious problems in the Navy. We were in the brown-water Navy when I was in command, at one time taking casualties at the rate of 6 percent a month, which meant that in any given year's tour your typical officer or sailor had nearly a three-quarters chance of being killed or wounded. I visited with 10 to 15 thousand of them in the Delta, and I visited with hundreds of the wounded in the hospitals, and I don't ever recall being taken to task about the war, but I recall many times being asked, "Why is it that I, who volunteered to join the Navy and volunteered to fight in Vietnam, am not permitted to look like my peers with a beard or a mustache or with sideburns?" It seemed to me, in looking into this needed revolution, that it was a small thing to give those patriotic individuals that privilege. Indeed, upon being reminded that Navy regulations had always authorized neatly trimmed beards and mustaches, it suddenly dawned on me that senior commanders were violating Navy regulations, and the Z-Gram in that regard merely said "Obey Navy Regulations."

Such things as beer in the barracks, black beauty aids in the Navy Exchange, the requirement that nobody be kept in line for longer than 15 minutes (instead of those very long lines that we used to have to wait through), were all designed to give our men and women, in the midst of a very unpopular war and surrounded by a hostile society, the feeling that they were needed and wanted and were the kind of individuals who deserved to be treated that way. And the bottom line, I think, is quite clear. By the end of four years our reenlistment rates were above 30 percent, nearly quadrupled.

MCPON Billy Sanders (editorial comment): *A positive attitude toward the Naval Service is so important. I don't believe that I have ever served with an officer who badmouthed the Naval Service and also was a success. Ninety-nine percent of them were poor performers, and they had serious flaws. The good officers take responsibility for the things they have to maintain and the orders they follow. I've heard junior officers say, "Well, the reason we've got to do it is because the captain says so." Or "The skipper wants us to stay late." They are not taking the attitude that the reason they are doing this is because the job needs to be done and they are making a decision to get it done. A few junior officers are poor leaders because they act embarrassed, as if the Naval Service is making them do something that they would not normally do. When an officer does that, he or she has lost the respect of the troops.*

Chapter 1. Study Guide Questions

1. What is the difference between a leader and a manager?

2. What is a leader's "service reputation"?

3. According to Admiral Uchida, how is an officer's humanity perceived?

4. According to Admiral Hayward, what ability must an individual have in order to be effective and successful in the Navy?

5. What does Admiral Zumwalt have to say about the element of surprise?

6. According to MCPON Sanders, how can an officer lose the respect of his or her troops?

Vocabulary

ultimate goal
proximate goal
entourage

coercive
social fabric
service reputation

Chapter 2. The Chain of Command

One of the most important elements of military organization is the chain of command. Adm. Teiji Nakamura sees the chain of command as

the fruit of the wisdom and experience of mankind which has been devised over a long period of time to help the services conduct complicated and extensive operations and/or administration effectively and systematically under a unified command. Today operations and management cannot be executed appropriately without this basic system.

In simple terms, the chain is the pyramid structure of communications, authority, and responsibilities that allows every individual in an organization to know what is going on with those below and what is expected by those above. It is the conduit for the orderly direction of command activities and provides a two-way communication flow. It is only as good as the people in it, all of whom are key links. Adm. James Holloway points out that

everyone involved in an operation must know what the senior is thinking and what his plans are. By sending orders down the chain of command, those involved make sure that the word is spread. By reporting up the chain of command, everyone between the person who is doing the job and the person who ordered it done knows what is being accomplished so they can adjust their actions accordingly. Uninformed individuals can take actions that are unsupportive and may even be in conflict with the operation. For this reason, the chain of command must be followed, and the chain itself must be well established; a tortuous chain of command that includes elements that do not need to know the information that is being passed indicates a deficiency on the part of the commander.

BYPASSING THE CHAIN OF COMMAND

The chain of command can be bypassed only when it is absolutely essential in the judgment of the individual who is going to bypass the chain that doing so is necessary for the national good. The apparent simplicity of this description does not do justice to the concept's crucial role in shipboard life. In addition to being a mechanism for promulgating orders and for "getting the word out," the chain of command is essential for maintaining good order and discipline, assessing combat readiness at each echelon of command, and delegating authority for action. Every officer in the Naval Service is an active participant in the chain from two aspects: from seniors downward and from subordinates upward. Each of these aspects encompasses different responsibilities.

Gen. P. X. Kelley observed that, during his first year as commandant of the Marine Corps, he was struck by the number of Marines who felt that they needed to go outside the chain of command to resolve a personal problem. In far too many cases, research had indicated that the problem could have been solved within the chain of command by Marines in positions of leadership and responsibility.

While General Kelley was addressing his deep concern over this subject at Headquarters, Marine Corps, the words of his distinguished predecessor, Gen. Wallace M. Greene, Jr., 23rd Commandant, were brought to his attention and were found to be as applicable today as they were in 1967.

General Kelley enjoined all commanders to give this White Letter the widest dissemination possible, and observed that, when individual Marines felt compelled to go outside the chain of command in the future, he would from time to time ask the question: "What actions were taken by the chain of command to solve the problem?"

A03C-jpp
3 Nov 1967

From: Commandant of the Marine Corps
To: Commanding Generals and Commanding Officers

Subj: Leadership and the Chain of Command

1. Since the founding of our Corps, continued success in war and peace has made the name Marine synonymous with soldierly virtue, military proficiency and professional pride. This legacy, passed on through generations of Marines, has been built on the cornerstones of camaraderie, discipline and esprit de corps.

2. Every Marine, from his first days in the Corps, is thoroughly indoctrinated in a way of life in which Marines take care of their own; one where each Marine may have complete reliance on his fellow Marine no matter the circumstance. One of the means of instilling this camaraderie and trust is to insure the opportunity to all Marines to express constructive criticism, grievances, and recommendations via the existing military chain of command, further to see that its availability is understood and exploited.

3. Discipline is not simply unhesitating obedience to orders. Certainly in the combat environment, response must be swift, sure and predictable. However, this requirement is not in conflict with suitable consideration of innovational recommendations prior to action, nor of professional criticism and review after. All Marines have both opportunity and responsibility to contribute to our administration, techniques and procedures. This is especially true with respect to matters which bear on the daily life, safety, and effectiveness of Marines. In part, it is through this process that we progressively improve. This rational dialogue is most useful and effective when it is conducted within the organizational structure.

4. Recently, there appears to be a growing tendency for some Marines to project problems, which could be resolved by internal Marine Corps measures, into avenues or to authorities outside of the Marine Corps chain of command. I do not mean to imply that any Marine is prohibited from communicating with anyone with whom he may desire. However, in almost all such cases, the query is eventually forwarded to the Commandant for response and action, whereas in reality most of these problems could be, and properly should be, resolved by a command level much closer to the situation than we are in Washington.

5. Paramount in the exercise of their duties, all commanders must continue to create, through traditional Marine Corps leadership, an atmosphere of faith, trust and reliance on one Marine for another; a feeling of esprit, protectiveness and interest which causes the junior, by choice and with confidence, to turn to his immediate senior for advice and assistance, assured that he may expect every reasonable effort to provide the appropriate information, counsel, action or remedy.

6. As I see it, we must positively impress upon all Marines that every commander in our Corps, from the Commandant to fire team leader, is vitally and genuinely interested in the health, welfare and professional ability of his Marines and that Marines of all ranks who have either a problem or constructive criticism to offer may best obtain action by utilizing the existing chain of command to express their thoughts. Utilization of the chain of command, in most instances, presents the most efficient and expeditious means for each marine's suggestions and concepts to receive the attention and consideration which they merit, toward the end that those concepts and suggestions which so merit are translated into action. We must insure a broader understanding that this is so.

WALLACE M. GREENE, JR.

The chain should never be so rigid that it inhibits imagination, innovation, flexibility, or good judgment. Situations will occur—and not necessarily only during combat—in which it is appropriate to bypass the normal chain of command and report an emergent situation directly to the executive officer or the commanding officer. Under such circumstances, an officer should inform his or her immediate seniors at the earliest opportunity of the situation that has arisen as well as why it was reported directly and what steps were taken to reconcile it. This is not done just out of respect for seniors; it also keeps key leaders of the command fully informed. Adm. Fran McKee concurs:

probably one of the better-established tenets of naval leadership is following the chain of command, and I think it's necessary that that be done as a general rule. There are certainly times, because of expedience or unavailability of an individual, when you can't do that, but that doesn't preclude going back after the fact to keep your seniors informed or your juniors informed. Everyone is part of this great wheel that's turning around to accomplish something, and if a couple of spokes don't know what's going on, then the wheel doesn't turn in proper balance and you're in trouble. So I think as a basic tenet the chain of command should be followed if at all possible.

Examples of circumstances in which an immediate senior might be bypassed are changes in the tactical situation or equipment casualties that might affect the ability of the command to meet its mission. Another exception would be an unsafe or an immoral situation that was deliberately being covered up by the next senior in the chain. Juniors are obligated to ensure that higher authority is apprised of such impropriety. Juniors must support the best interests of their commands and uphold the moral obligations of their commission. This may require making a request that an immediate senior pass critical information up the chain of command. If such recommendations are unsuccessful in bringing about a resolution, an officer may be obligated to bypass the chain. Fortunately, situations of this sort are rare. Nevertheless, junior officers are expected to place the safety and operational readiness of their commands above loyalty to an immediate senior if there is a question of negligence or impropriety.

Gen. Louis Wilson advises that

if someone is given an order which he believes to be morally indefensible, he has every right to bypass the chain of command. I cannot help but think of the My-Lai instance in Vietnam, where indiscriminate killing of civilians was ordered, and it has not been recorded that there was any protest. Now that's very easy in combat, and we must be careful not to condemn people for this, because this is a fine line.

I'll give you an example of an experience that I had in the past. We went into a Vietnamese village in which there was nothing but women and children, all the men had gone, and as we left the village, five of our men were killed—fired at from holes in the ground which are generally near the trees. The next day we went by, and suddenly there was fire coming from the holes, and we went in, lost a few men, and there was no one in the town but women and children. There were these tunnels in Vietnam by the hundreds through which the enemy could move about.

There's a point in time when a commander has a moral obligation to the parents of sons under his command, and he cannot afford to continue to take those casualties, yet on the other hand he knows that if he directs that artillery fire or air strike on the town, he's going to kill women and children. Now, what do you do? What do you do, Lieutenant Jones, in this instance? Is this morally defensible or not? I think it's a judgment that one has to make, and you can be criticized in the papers either way, for being a poor commander to take casualties and not take care of your men and for wantonly disregarding the lives of American soldiers, or, on the other hand, for killing women and children.

I have to say to you there is no answer. I'm not here to tell you that it is black and white, and I don't believe anyone else can. It's a situation that calls for good judgment, and those decisions have to be made. But when you make the decision, then you've got to live with it.

If you say go ahead and bomb the village, very well, but there may be a lieutenant who says, "I cannot go along with that, and I hereby submit my resignation," and he leaves the scene and goes back to division headquarters and says, "Not only am I submitting my resignation from the service, but I'm going to the newspapers and I'm going to report this." Therein lies the problem for each person. Each person, though, cannot fight the war himself.

These are the hazards of command. If you would be a commander, you have this responsibility you have to assume yourself. You cannot push it up and you cannot push it down, and if you're not willing to live with those decisions, then you ought to seek another profession.

From time to time, seniors also may direct action to a junior who is not directly below them in the chain. Similarly, on occasion, a senior may convey wishes or important information in informal situations such as social events. In doing so, as a matter of convenience or practical necessity, a senior expects that the person receiving such information or direction will subsequently convey it to interested parties. The chain provides a system of checks and balances. If key people are bypassed, the guidance they normally provide may be lost. Most important, someone in the chain may know why such an action should not be undertaken.

Admiral Nakamura believes that

it is a basic principle that military personnel give orders and instructions and report or offer opinions through the chain of command. The violation of this principle invites confusion and exerts a negative influence on leadership. But it is permissible to bypass in the following cases:

1. When it is perceived that crucial timing will be interfered with by strict observation of the chain of command.
2. When it is necessary and suitable to bypass in order to simplify communications on routine affairs.
3. When seniors direct a bypass, recognizing a special necessity because of security or other reasons.

When the chain of command is bypassed, the commanders bypassed should be informed concurrently except in the case of (3) above. When it is not possible to inform them concurrently, they should be informed as soon as possible.

Adm. Paul de Bigault de Cazanove believes that

it is important that the chain of command never interfere with proper operations. Orders are to be clear and simple and understood by everyone in the chain, for bypassing will interfere with properly coordinated operations. Sometimes it may be easier, for more-junior officers, to go straight to the person detailed to do the job instead of passing through the proper petty officer in the chain of command. This is not to be done, because the individual who is bypassed will just be inclined not to do anything further when called upon at some future time. You cannot tell someone he is not important at one point of time and then that he is important at another point of time. It must be clear that everyone has a job to do and that each person has to get the order from the one he is supposed to get it from, and not from someone else. Of course, in emergency situations it may be necessary to bypass the chain of command, but in that case, those bypassed should be informed as soon as possible as to what transpired.

THE IMPORTANCE OF ACCURATE INFORMATION

When faced with a difficult decision, a commanding officer is reassured by the knowledge that the information given him or her accurately reflects the existing situation. In a combat situation, in which the captain's decisions must be based on a rapid assimilation of information, survival may depend on accuracy. But accuracy is equally important during daily routines, as the commanding officer evaluates the morale, climate, and combat readiness of the ship and crew. A critical input comes from the offi-

cers and petty officers in the chain. The responsibility of the commanding officer to know his or her command makes good communications with subordinates essential.

Senior officers should exercise control through policy rather than through detailed procedures, for it is important to allow a subordinate to handle the details of a specific task if he or she is capable of doing so. An example of this can be found in the *Marine Officer's Guide.* Remember the old promotion-examination question for lieutenants, in which the student is told that he or she has a ten-person working party, headed by a sergeant, and must erect a 75-foot flagpole on the post parade ground. Problem: How to do it. Every student who worked out the precise calculations of stress, tackle, and gear, no matter how accurately, is graded wrong. The desired answer is simple. The lieutenant turns to the sergeant and says, "Sergeant, put up that flagpole."

Under normal circumstances, formal dialogue takes place primarily between the commanding officer and the executive officer regarding administrative matters, between the commanding officer and the department heads regarding operational and administrative matters, and between the commanding officer and officer of the deck (conning officer) or tactical action officer regarding the tactical disposition of the ship. In every case, those briefing the commanding officer must rely on their experiences and the information they have obtained from subordinates to advise the captain correctly. Thus, the chain depends on a series of valid reports.

These reports must include the bad news as well as the good. Sometimes junior officers make the mistake of believing they are expected to solve all problems themselves. Therefore, they do not want to inform their seniors of a problem until it is under control. While it is better, as the old saying goes, "to come with a solution rather than be part of the problem," it is both an obligation and common sense to bring those with greater experience into the problem early. In addition to familiarity with past problems and their solutions, seniors usually have greater access to resources and the clout to generate necessary assistance. More important, seniors must be aware of weak areas in their commands in order to direct the command's activities effectively. When in doubt, it is better to err on the side of keeping seniors fully informed in a timely manner, especially when conditions are deteriorating.

Junior officers must have the courage of their convictions and inform a senior when the senior's perceptions are believed to be in error. Candor and honesty are essential at all times. This is not, however, a license to question authority, to lack tact, to be inappropriately stubborn, or to step outside the bounds of discretion. Fortunately, most senior officers can readily determine whether subordinates are speaking their mind or are inhibited and only saying what they believe the boss wants to hear. The latter types serve no one well. A fighting team needs to have everyone thinking, communicating, contributing, and cooperating. One trait of American naval officers that many of our foreign colleagues admire is the readiness of juniors to "call it like it is" and to speak up when they believe there is a better way. The trust, confidence, and instant responsiveness of the chain of command that is required in emergency situations is not meant to foster blind obedience.

Sometimes, limited experience and lack of confidence cause newly commissioned officers to rely too much on their chiefs and leading petty officers for initial solutions to problems and to transmit these recommendations to their seniors (department heads) without supplying their own input. No matter how inexperienced, an officer should not underestimate his or her importance in the chain. A junior must always apply common sense, dig until a problem is understood, and give a personal best judgment. Constant interaction with seniors and subordinates in the chain will quickly broaden experience and enhance knowledge.

Maintaining the Integrity of the Chain of Command

We frequently speak of the integrity of a system, usually meaning the ability of a mechanical or electronic system to function reliably under a wide range of conditions and stresses. This is equally true for the communication and control system defined by the chain of command. Within military organizations, the integrity of the chain of command is a direct reflection of the character of the officers and sailors or soldiers who comprise the organization. Where the integrity of even one officer or petty officer within the active link is weak—whether through lack of moral courage, a self-serving attitude, or limited understanding—the system will deteriorate. Officers in authority will not have an accurate assessment of the capabilities and limitations of their command or the benefit of the best judgment of their subordinates. Consequently, the success and morale of the unit could suffer.

When the chain of command is not used properly—when the chain is bypassed downward under normal circumstances—there may be several detrimental results. First, the bypassed persons will feel that they do not have the confidence of their senior, and they may harbor resentment, up and down. Second, it may make them hesitant to act in a similar situation in the future. ("Didn't the boss handle that without reference to me last time?" they may ask themselves.) Third, a senior officer does not help subordinates by doing their work for them. They can be instructed and guided as necessary, but they should never be cut out of the chain of command. Finally, seniors should be training themselves for higher, not lower, command.

An active chain of command should also ensure that everyone knows his or her place within the organization and that individual efforts contribute to achieving the overall goals of the command team. A shipboard division in which the members receive conflicting commands from their leading petty officer, their chief petty officer, and their division officer cannot be managed effectively. Similarly, contradictory priorities from the commanding officer, the executive officer, and the department head breed confusion. It is imperative, therefore, that priorities originate from a single source (however many centers of emphasis there may be) and that each individual feels personally responsible for his or her contribution to the overall success of the command.

Span of Control

Span of control means that there is a limit to the effective supervisory capability of any one individual. The span of control typically is three to seven people, though it may be higher. The following illustrates the importance of the chain of command. A rifle platoon commander has 30 to 50 subordinates, a company commander has between 150 and 200 people, and on up, to those who have several million people under their jurisdiction. All of these officers must directly supervise a number of people, each of whom in turn supervises several more.

Secretary of the Navy James Webb suggests that

whenever possible, a leader should choose people to work for him whose strengths compensate for his own weaknesses. These people are in diverse ways mirrors of different parts of that leader's personality and professional acumen. In turn, the leader should allow those he supervises to choose the people they want to work for them. The person in charge may want the final say in whether an individual is offered a position, but the person who will be supervising that individual knows best whether he can work with him or not. [Editor's note: Of course such leeway is not always possible, and sometimes it is not desirable—for example, if it would result in significant friction between members of subordinate supporting units.]

Once he has selected his staff, the leader should

trust them fully to perform their responsibilities and see that they are rewarded for doing so.

Rewards can take different forms, but, as Napoleon said, the general's reward is not a bigger tent, but command; in other words, the ornaments of leadership should not override the responsibilities.

THE CHAIN AS A TWO-WAY CONDUIT FOR INFORMATION

When the chain of command works, it is an admirable institution. Secretary Webb recalls how it worked in Vietnam.

I got to Vietnam, and the very first patrol I went out on, I said as we were walking along something like, "Hold it up," and a column stopped, and I said, "Get on line," and they were on line. I mean almost before I could get the words out, that's how fast it was, and it was because they used the chain of command. I would say something and the squad leaders would say it, and the team leaders would repeat it, and it was done, I mean within seconds. It was marvelous.

A senior must take time to pass the word down the chain. Sometimes a senior's initial concerns are satisfied, but juniors receive insufficient information about why, ultimately, certain decisions were made or about what is going to happen. Juniors deserve the courtesy of being kept well informed. In most instances, the crew should be told why policies are adopted; all parties who have responsibility or could help with a problem need to hear about it. It is also easier for them to be team players when the objectives are understood.

It is also important to consider at what level in the chain various decisions should be made. Some issues cannot be delegated. However, giving authority to more junior officers and petty officers often enhances efficiency, teamwork, and a sense of responsibility.

A commanding officer or department head may deliberately probe into what is going on at the "deckplate level." Wise commanding officers will not totally depend on what comes to them directly through the chain of command and must be alert to "filtering" of information by subordinates. Commanding officers are encouraged to talk to their people, to meet with their chiefs, to tour their ships frequently, to hold captain's calls with the entire crew, and to keep their eyes and ears open. At the same time, juniors can be assured that commanding officers will keep informal comments in perspective, that they know their people, and that they will avoid undercutting the officers and senior petty officers in their own chains of command. Tasking should flow through the chain.

MCPON Sanders: *We tend to have a better chain of command going up than we do going down. I think if we kept the sailors better informed of what is happening in their organization, it would make them feel part of the organization instead of being just workers. They are contributors and contributing because they want to. If they are kept informed, they will know why something has to be done without having to ask or wonder why. While I would agree with the concept that unless emergency situations dictate otherwise, the chain of command should be followed, I do think that sometimes some of our junior officers use the concept as a crutch by waiting for the chain of command to make a decision and get the job under way, as opposed to taking some initiative and responsibility and starting the job. Follow the chain of command, but don't use it as an excuse to put off directing that necessary action be taken.*

Chapter 2. Study Guide Questions

1. According to Admiral Nakamura, why is the chain of command so important?
2. According to Admiral Holloway, what is the benefit of following the chain of command?
3. When should the chain of command be bypassed?
4. What actions should be taken when it has been necessary to bypass the chain of command?
5. What must a junior do when a senior's perceptions are believed to be in error?
6. What are the possible detrimental results that may occur when the chain of command is bypassed?
7. What is the typical effective span of control?
8. What is MCPON Sanders's advice regarding the use of the chain of command?

Vocabulary

chain of command	reconcile
pyramid structure	perceptions
cornerstone	candor
dialogue	span of control

Chapter 3. Standing by Seniors and Associates

"Authority and responsibility is invested in the captain of a ship, and the unequivocal requirement is made for obedience to his command." The concept embodied in this quote serves as the basis for the following discussion of the naval officer's obligation to support both seniors and associates in their policies and actions.

A fundamental assumption all leaders must make, regardless of their station in the chain of command, is that every leader above them is morally motivated to carry out the mission and, whenever and wherever possible, to carry it out in a manner that serves the best interests of the officers and enlisted who make up the Naval Service. A professional tasking that does not spring from this premise will probably result in an undertaking that falls short of the intended goal in some respect. Guided by this assumption, officers comprehend that their response to commands, orders, and other official taskings from higher authority, no matter how challenging they are or how many work-hours they require, must be rapid and unquestioning. This is the essence of military discipline.

Most dramatically, an officer's instant and whole-hearted response to commands in combat or in critical damage-control situations is passed rapidly down through the ranks and can result in saving a great number of lives or a ship, submarine, or aircraft. There is a great difference between leadership and command. "A leader is a commander, but a commander is not necessarily a leader." People can be given a command by an official order, but they cannot be given the qualities of a leader. They are leaders only by the use of their own qualities. In a less dramatic sense, a like failure to respond with utmost vigor, aggressiveness, and professionalism to a firing exercise, a man-overboard drill, or a flight-deck barricade drill could well cost a ship and the crew the Battle "E" after arduous months of planning, training, and striving for that ultimate operational recognition. The Battle "E," incidentally, is an award that every ship or squadron must set its sights on during each competitive cycle if, to paraphrase a popular recruiting slogan, a command is going to be "all that it can be."

DEMONSTRATING LOYALTY

Recognizing the motivation and intent of officers in command is part of a precept of good leadership: loyalty. The last thing a good leader wants or needs is a "blindly" loyal subordinate, but an informed and loyal subordinate is highly valued. Loyalty works both ways. Loyalty to one's own organization and personnel is just as vital as loyalty to seniors. If a leader is proud of subordinates, if he or she has the faith in them that real loyalty demands, they will return such support and backing a hundredfold.

An interesting exposition on the subject of loyalty is provided by Adm. Karl Clausen.

Loyalty is one of the most noble qualifications that can be held by humans. An officer owes his seniors and his country loyalty. In this lies some danger, because there will always be those who try to take advantage of it. This and history explain why, in the German Bundeswehr, all the particular aspects of loyalty have been defined by law [Soldatengesetz], in order to make plain the limits, which must not be passed over. The following conditions should be known to all officers:

1. The law must describe exactly the manifold virtues specified in the catalogue of duties.
2. In the forces it must be clear: Although a catalogue of duties is demanded from everyone, especially from officers, these duties and virtues cannot be demanded unconditionally. They have to be derived from principles and fundamental rules of constitution and existing law.
3. Officers must be brought up with a critical mentality. Above all, they should be conscious of intentional and unintentional possibilities of misleading. That implies that they must be informed on civics.

The education of German officers before the war is a warning example. They were educated according to a law of 18 June 1921 [Wehrgesetz], which said: All members of the military service (officers, soldiers, seamen) are not allowed to participate or take an active part in politics; are not allowed to participate in any kind of political election; are not allowed to participate in political parties and political events, meetings, or assemblies; and cannot be elected into a political function. This law was made in the time of the young German democracy, 1921, as a guide especially for officers. I was brought up in accordance with this law very correctly. The law was cancelled, to my knowledge, in 1944. Its results and effects were fatal.

DISOBEYING ORDERS

The following statements by Adm. Sir Nicholas Hunt and Adm. Arleigh Burke should be viewed in light of the fact that the comments and recommendations were made by very experienced officers. Admiral Hunt provides guidance from his position as leader of the British Fleet.

It is very important—increasingly so as an officer gains in seniority—that he understands what responsibility for his actions means. The more independent his position, the more he must be prepared to regard his orders as nothing more than guidelines. I call the process of learning to judge when this applies "training to disobey." The history of war gives us many well known cases of great successes being achieved by those on the spot who have followed their own instinct in contradiction of their orders; and equally of failures or disasters which have followed from slavish obedience to orders which were no longer—or never were—appropriate. Battles have been won and lost on this account alone.

If an officer disobeys his orders, and proves to be right, he should be promoted. If he does so, and is wrong, he should be ready to take responsibility for his actions. But those who then decide his fate should distinguish between energetic initiative (which may yet be an invaluable asset for the future) and the characteristic of poor judgment (which may be a grave liability).

Admiral Burke says that

in combat, a leader must always think in the enemy's terms. When he thinks as he imagines the enemy is thinking, he can often discover how to prevent the enemy from doing what he ought to do. The enemy will also try to think like the U.S. officer, so the U.S. officer sometimes must do the unexpected.

In going over some battles with Admiral Kuzoka, a Japanese commander, I found that he knew more about what I did, about our tactics, than I did, because he studied it. He had his charts, photographs, all the other data, and he knew exactly what we did. But he one time said, "You did not follow your own doctrine. How could you operate and not follow your doctrine?" It was a doctrine, it was not an order. It was a normal thing to do, the most logical thing to do. It had been studied very carefully. "But," I said, "I knew you had our doctrine, you must have, and I figured that you probably had it, so I didn't do what the doctrine said. I did something else, I split my forces and arranged a big surprise." He said, "How can you disobey orders?" I said, "Those are not orders." The Japanese were rigid in their operations, and we were not.

These views of senior officers are provided so that junior officers will understand the complexities of command that must be considered by their seniors and thus will be able to follow orders loyally by having confidence that their seniors' orders are designed to win battles and minimize loss of life. On this same subject, Adm. Robert Long comments that

if decisions made are contrary to an individual's personal beliefs, he has several acceptable options available to him. First, and the one most recommended, is for the individual to enunciate the reasons why he thinks the decision or the policy is not correct; he should always recommend an alternate course of action. If the issue is of great magnitude and an officer cannot in all good conscience carry out the decision and is unable to change it, then he clearly has the option of resigning or retiring.

Back in my own life in the Navy, there were times when I was sorely tempted to turn in my chit and resign because I felt so keenly. Most of those issues dealt with decisions of seniors affecting my own family life, and also decisions that affected the welfare and morale of the people under my command. In those instances I did object to the policies and decisions that were being made, and when those decisions were not changed, I must admit that I was tempted at that time to turn in my resignation. As a matter of fact, at one time I did write out my resignation and was argued out of it by my wife, and on reflection, I'm glad she did.

Every officer will be faced with difficult decisions during his or her career in the military. But no one should take a drastic action without a great deal of thought and consultation with others.

Gen. Robert Barrow gives two examples of situations in which he deliberately disobeyed orders he disagreed with.

I happened to belong to the school that says there are some circumstances in which an officer can refuse to do what he is ordered to do. There are some people, surprisingly, who say you are to execute whatever it is you are told to do without question, even though it is distasteful and you don't want to do it; they say it is your duty to do it, etc. I'd say normally that's true, but, this world being made up of imperfect people, you're going to have people give you wrong orders. Just to oversimplify it, if someone said, "I want you, Captain Barrow, to take your company and jump over the cliff," obviously I would not obey. But it doesn't really have to be that blatantly stupid, it can be something much more subtle than that.

When I was involved in the Seoul [Korea] operations, for example, I had a company that was out in front of all the other companies in an irregularly shaped line in the heart of Seoul. I had an eerie feeling that something was not quite what it should be; the area that we were going into just looked dangerous. There was this intuitive thing that said don't go blindly charging ahead because something's wrong down there. So we very carefully positioned ourselves, sought proper cover and protected it, and we watched with binoculars the area below us and in our zone of action, and all of a sudden we saw significant movement, enemy movement, back and forth, repositioning themselves to give us a welcome party in many, many greater numbers than us. Had we gone into this, we would have been going into a real trap.

It was an ideal situation to employ a support in arms, we wouldn't expose a single Marine on the ground to enemy fire, but we would blast the enemy with artillery, mortars, long-range water-cooled machine guns, and anything else we could think of. And that's what we did, we brought up all these forward observers and controllers of these kind of weapons and started pouring in on them, and as we did they, too, had to seek better cover and expose more and more of their presence. Meanwhile my boss (I'm a captain, my boss is a lieutenant colonel) has me on the radio saying, "Move out," and I'm trying to tell him that what I'm doing is a lot more important than me saying "Everybody move out," which would get a lot of people killed. I told him what we were doing was accomplishing a mission in the sense that we were killing the enemy and were going to be able to go into that area without ourselves getting seriously hurt. He implied such things to me as, "If you don't move out at once, you may have a new battalion commander," which was his way of saying, "They are really putting the heat on me for you to move out."

I thought I was going to go nowhere, so I turned my radio off, which is a form of disobedience, and I got my most articulate platoon leader, the guy who was persuasive and forceful, who was an effective speaker, and I didn't need to tell him what the circumstances were, he happened to be the platoon leader of the third platoon, he could see for himself. I said, "Go back there and find the command post or the battalion and entreat our commanding officer to come up here and see for himself." I said, "I'll leave it to you to persuade him. Tell him what we're doing." Lo and behold, he did just that, he convinced him that what we were doing was the right thing, more important than obeying someone's orders.

I may as well digress here and say that the orders to move out came from way up the line and had to do with orders we got that night about

moving out, all beholden to someone's idea that Seoul should be retaken on the third month anniversary of its falling. If there was ever a stupid thing, it was to make a military decision based on some anniversary. There are no military considerations in that at all, but that's really what somebody was doing. Some general way up the line was saying, "Let's give MacArthur a prize. Three months to the day that it fell, we're going to take it back, so move out Marines, move out," and you know Marines are very obedient, so it's easy for a regimental commander to say move out when he hasn't seen any reason not to, and then the battalion commander the same thing, but when it got down to me, I didn't blindly say, "Yes, Sir," because I had something else going on, and I balked. I didn't move out, and that's what you get paid for. You see, there are people who would have moved out when he said, "I'm telling you to move out, attack, go forward," and there are many Marines and others who would say, "I've got my orders. We've got to do it." But I believe it is a matter of judgment. Now this is a touchy business. When do you not do something you're told to do and be prepared to take the consequences? Suppose he came up there and said, "God damn it, I told you to move out, and I don't give a damn what you're doing, you're relieved." I would have gone to my dying day thinking I did the right thing, but he could do what he wanted to, get rid of me if he wanted to.

To complete the story, the lieutenant brought him back and we were then almost in the anticlimactic part of what we were doing without supporting arms. We were about to cease, and yet there was sufficient going on that he got excited and jumped up and down and said, "Let's get more artillery, let's bring in more mortar," and we almost felt like saying, "Well, for God's sake, if you could have just been here about an hour ago, you would have really seen what it was we had as targets." So I'll just say I survived that, I didn't get any criticism or anything else. I've done that sort of thing several times in my career, simply not doing what someone told me to do, and I had to maybe sometimes stand there with my heels together, and if I believed I was right, that's the way it was.

About one time I will never know whether I was right or not. This same battalion commander had a good reputation for being a fair man, he was a prisoner of war during World War II and escaped to the Philippines. He was a Naval Academy graduate and had a good record. He took over the battalion in Camp Lejeune, this was 1949, the year before Korea, and I had just taken over a rifle company about the same time that was at Little Creek, Virginia, so we were miles apart and had never met. I didn't know that the first thing he did after he took over was to say he wanted to go up to Little Creek and see this company that I happened to be in command of. My company was involved in the evaluation of submarines carrying troops and an amphibious tractor and other things, and we'd go under and surface and do all kinds of things to see how quickly we could move x numbers of people. We did many tests and drills. It was a group that was involved in something different.

One day we were not doing any of this evaluation, it was sort of a clean-up everything day. A lieutenant colonel from the headquarters that I was responsible to, a staff officer, had sent a classmate of mine, a captain, to tell me that another unit, a battalion-sized unit, Marine Corps, had moved out of some barracks not far from where we were and left it in a horrible state of police and that he was directing that Captain Barrow and his A Company, 1st Battalion, 2d Marines, go over there and clean it up. That's wrong. I mean, that's wrong on his part to do that. It required an outfit that, it looked to me (I had just gotten there), was pretty squared away and had its own area cleaned to go clean up someone else's mess. So I balked at that. I told the captain, "You go back and tell the colonel I'm not going to do it. We didn't make the mess, also we're not going to do it." He came back in about twenty minutes and he said, "Man, the colonel is furious. You've got to go up there and see him."

What would have been the outcome normally? Would I have gone there and been chewed out and still have had to do it, or would I have been able to talk him out of it? Would I as a young captain saying, "Sir, I should like to come on out and talk about this, there's a morale issue involved," have been able to talk him out of it? Well, as it turns out, as I was waiting outside his office, down the hallway came this lieutenant colonel with another officer I knew with him, and he stopped and said, "Colonel Hawkins, this is one of your company commanders, Captain Barrow." The battalion commander, as I said earlier, had come up to see us in our training. We hadn't seen him, but he had just arrived, maybe within the hour. He greeted me cordially and I did him respectfully, and he said, "Everything all right? You got any problems or anything?" and I said, "Yes, Sir. I have a problem." So I told him about this incident, which was proper, since he was my battalion commander and someone senior to me out of my chain was about to pull my

chain. He didn't get angry, but he said, "Humph, Captain, I'll take care of that." It turns out that he was a classmate of this lieutenant colonel who was after me, but he was also senior to him, and I'm going to assume he went in there and said to him, "You know that's not the right thing to do, making that outfit go over there and clean up somebody else's mess. Why don't you make sure they send a contingent down here to do it or make some issue out of it, but don't go and do that." I don't know what he said, but the point is, I didn't have to face that music.

Now think about this story. Everything has transpired within the earshot and eyesight of several people, my officers, first sergeant, company gunnery sergeant, clerks, and whoever else is around close enough, and their antennas are quivering. "This young new captain we've got has done something none of us ever heard of before. He told another captain to tell the colonel he wasn't going to do something. What in the hell do we have here?" They knew what the task was and "By God, this young captain ain't going to make us go over there and clean up somebody else's mess. He has said he wasn't going to do it, and then a little while later he's been sent for. We're going to have to do it, but he sure is standing up to them." I came back and never said anything, and I would guess, being a judge of human nature, that my stock went up considerably. I didn't come back making any claims about anything, I certainly didn't tell any stories, but I didn't want to say, "My butt was saved. I ran into the battalion commander and he took care of it." I didn't need to say it, so I didn't say it. Maybe I exploited the situation for my own benefit. But that's not why I did it. I did it as a matter of principle.

You can just imagine what those young Marines were saying about that new captain. "Here's a guy who stood up to them. Not only that, but they sent for him, he came back, and we still didn't have to do it, and it looks like he's in one piece and he's still got his job." That was an example of standing up for principle, even though we might have had to do it. An additional point is that the good consequences of standing tall on morale issues can sometimes be of marvelous value to a commander. You do it because it's right, but you also end up getting a followership that becomes even more intense.

On the subject of a junior advising a senior about disagreement with policy, Secretary Webb stated that

many people benefited when I spoke up to correct

a wrong. In Vietnam, we had an area called Liberty Bridge, and eight miles away there was this place called An Hoa, which was a combat base, and there was a dirt road that connected the two where a convoy ran twice a day. Because the dirt road was occasionally mined, every morning two rifle platoons whose duty rotated provided flank security out in the weeds while the engineers swept the road for major mines. But you know the Vietnamese, the VC, were not stupid, and they realized that if they put a mine inside the road it was going to be found, so they started booby trapping both sides of the road, and we averaged a loss of a man per day in this small platoon that averaged 28 people on the flanks.

I had been in Vietnam not very long, maybe a month or so, when my platoon got this mine-sweeping flanks security. These guys were always saying, "Well, you know somebody's going to get it today, it happens all the time." My people moved along and I lost a guy in the right flank the first day, and I came back and I started talking to my company commander about us and I said, "It's really silly, you don't lose anybody on the road and yet in the flank security, for the guys on the road, you're always losing people," and I went out the next day and I lost two guys on the left-hand side, then I went to the battalion commander, and I said, "This is crazy, you know, count the casualties that you've had, Sir, and you know for the sake of possibly losing somebody on the road, we're guaranteeing three and four hundred people a year out on the flanks," and he had me write a proposal, which was rather unusual in the field of Vietnam, but I got together with my company commander and we wrote a proposal and we changed the policy so that we were allowed to send patrols out and set up sort of outpost positions on the side of the road, and the casualties stopped.

Good officers who believe strongly that something is wrong have the courage to say they will not do it, and generally they will be listened to, although at times they will not.

The officer who believes he has to present himself in different ways to different people will end up wrapped around the axle. If he understands himself and is able to live with himself, then the officer only has to say it one way, and that makes life much simpler. The officer who adheres to his value system generally is at peace with himself and is willing to take the consequences for his beliefs and tenacity.

MAKING UNPOPULAR ORDERS PALATABLE

The Naval Service is becoming increasingly ex-

pensive and sophisticated, both in terms of combat and engineering systems and in terms of the ever better educated and trained officers and troops who provide the vital human link in the war-fighting equation. Leadership in the modern Naval Service is, more often than not, beyond the capacity of any single human being. Good and wise leaders at each level in the chain of command, therefore, are receptive to constructive and knowledgeable inputs from subordinates when the situation warrants. Junior officers should express dissent to their seniors in private and in a frank and full manner. They should never be open in their dissent or air "dirty laundry." Of course, after discussing the matter with their seniors, they must faithfully and loyally execute an order and impart full backing of the policy to subordinates, whether in agreement or not. Effective leaders will readily accept and weigh such inputs with knowledge already at hand, and then make their decision and issue appropriate direction and guidance.

The direction or guidance handed down may not be what the subordinate leader recommended or hoped to influence through constructive input, but this is where "the rubber meets the road," where loyalty comes firmly into play. An officer who sets the example of loose criticism, especially in the presence of enlisted, cannot be surprised if he or she gets only a similar lack of loyalty from the subordinates. The subordinate leader is duty bound to run with this order as if it were his or her own and to "make it happen!" When an officer receives direction from a senior officer which he or she considers wrong or unfair, or knows will be less than favorably received down in the ranks, the junior officer must confidently draw from a reservoir of leadership "tricks" and from the well of rapport that has been established with subordinates to get the job done properly and in good time.

While commands must be carried out in a prescribed manner, and Standard Operating Procedures are to be rigidly adhered to unless operational necessity dictates otherwise, subordinate leaders, more often than not, will have the option of carrying out orders or other taskings in the manner they deem to be in the best interests of the Naval Service. An imaginative and resourceful leader can often soften the overall impact of an unwelcome order or task.

EARNING THE LOYALTY OF SUBORDINATES

Adm. Red Ramage, in talking about training troops for combat and following orders, noted that

there may be an occasion to question the policies of your seniors, and, if so, you must be ready to offer substantial reasons therefore and similarly listen to subordinates if they question your orders. If unable to settle the differences, you have to seek further advice and counsel; in other words, there are times when you say, well, that's not going to work, and I did this at times with Admiral Felt when he was vice chief and he would have everything all prepared. He'd studied all the references to some problem and he'd made up his mind the action that should be taken and he called in all of his staff officers and said this is the way it is going to be, and I'd say, well, I don't think so, I think we should approach it a little differently and so on, and he'd listen and say, well, I think you're right, and many times he'd say, you're the only guy that ever takes exception. Well, it can be done and seniors will respect you, and likewise, in all my commands I had staff conferences and I conducted them almost like a TV show, I'd ask the individuals questions, throw the problem out, let them respond and get their input, because nothing succeeds like success. If staff officers find that their suggestions or recommendations are accepted and acted on, it increases their ego and their ability and eagerness to participate further.

To this, the interviewer, Prof. Karel Montor, said: "It sounds like what in effect you're doing is training. You're building in subordinates a responsibility to think, probably knowing that sometimes they may have said something to you which you didn't agree with, but the fact that you didn't hit them over the head and say 'Get out of here' in turn made them realize that they could have confidence in saying what they thought and you'd listen, even if you wouldn't necessarily always accept what they had to say."

Admiral Ramage continued:

Well, the only way you're going to build the strength of your command is by enhancing the responsibilities of all the individuals associated therein. I carried this a step further when I had command of the First Fleet. I was very much concerned about all these vast operations orders we used to have in the amphibious force, with each level of command issuing their own supplementary operation orders covering every last detail. They became so voluminous they became confusing and self-defeating. I often compared them to Shakespearean plays where every actor knew ahead of time when he was going to be on stage, what his lines were, and when he would be off stage just standing by. It was very unrealistic, to say the least. So I called in Admiral Colwell,

ComPhibPac, and General Krulak, CGFMFPac, and told them we were going to run this next amphibious exercise without an operation order. Instead, a Letter of Instruction (LOI) would be issued giving the Task Force Organization, task assignments and time of getting underway. From that point on the exercise would be directed by dispatches in accordance with a scenario prepared in advance by my staff. Ships would assume full readiness conditions upon getting underway and assume actual wartime conditions. That is, minesweepers would have to proceed ahead and sweep for mines already laid in advance. Then they would have to take suitable antisubmarine measures to avoid submarines awaiting their sortie. Next, they would have to man their anti-aircraft batteries and take targets being towed by planes under fire. And eventually the surface support ships would be turned loose to conduct shore bombardment on San Clemente Island. During the night, ships had to be on the lookout for surface raiders exiting from San Clemente. Also, an active air search had to be maintained as occasional bogies would make unscheduled attacks throughout the night. Every effort was made to simulate actual wartime conditions and to put the onus on each command to react to each situation in accordance with doctrine.

Specific directives were sent to the Task Force Commanders requiring them to initiate certain actions. This in turn would result in further directives being originated and passed on down the line, calling for individual initiative and imagination at each echelon of command. This put a tremendous premium on communications and served to impress everyone with the importance of this essential adjunct to successful operations. Many other innovations were injected into the scenario. For example, up at Camp Pendleton, where the landings were made, the Marines had tried to simulate conditions to be found in Vietnam as closely as possible. They had built several native villages—no one was allowed to speak English—only Spanish was permitted. The natives were not too cooperative, and it took quite a bit of ingenuity to accomplish necessary tasks. Everyone entered into the spirit of the enterprise and came up with many original contributions.

This type of operation was so successful and so realistic that it was followed in all subsequent fleet exercises. In each one, special emphasis was directed at some particular aspect of operations, such as: communications, guided missiles, ECM/ECCM, NTDS, AAW, ASW, and strike warfare.

All facilities both ashore and afloat were utilized in order to provide all realism possible.

"Loyalty is in no way merely a blind and servile service to the letter of the regulations." One officer remembered a story that illustrates this. He had to tell his carrier's crew, during transit home from an arduous seven-month deployment, that the type commander had informed him that his staff had determined that preparation for a unique, out-of-home-port, service-life-extension overhaul, scheduled to begin six weeks after the ship's arrival home, would require ten hours of ship's force effort every day of the six weeks at home, including the day of arrival:

The fact that it was summer and married crew members' children were out of school and looking forward to spending time with their long-absent fathers, and that the local beaches were crowded with distractions for the unmarried crew members, certainly did not make my 1MC (public address) announcement any easier. I followed up the news with a challenge to the crew to figure out a better way to execute the tasking than the type commander's staff had formulated, and we would do it our way. Needless to say, the minds of my 3,500 crew members went into overdrive, and the results were even more impressive than I had imagined: first of all, we were able to carry on with the normal post-deployment 50 percent leave policy during our first 30 days at home; second, by a huge consensus, the crew elected to go with a unique 0700 to 1300 workday, forgoing lunch and other workday-associated personal needs that would normally be satisfied during a standard workday, until after the 1300 secure.

In doing it our way, we were able to satisfy the type commander's requirements while coincidently ensuring that the best interests of the crew and their families, and, in the long run, the best interests of the Navy, were served.

Admiral Long discussed how priorities differ in peacetime and during times of war.

One of the principles of command says that the mission comes first and the welfare of our people comes second. I do subscribe to that view in critical wartime situations. Obviously, the situation or the scenario dictates the priority of what we are doing, and military officers need to have sufficient flexibility in their own thinking and decision making to accommodate varying priorities. In wartime, where the accomplishment of a mission is critical, I would certainly subscribe to the view that that mission is of primary importance.

However, in peacetime, one of the principal jobs military officers have is to maintain the readiness of the force. The readiness of the force clearly starts with personnel readiness, and that involves retention of trained, experienced personnel, and that in turn is dependent upon the welfare and morale of individual people. Therefore, an officer in a command position must, in my view, in peacetime always be concerned about the welfare, the morale, the training of his own people as necessary, because that concern is absolutely fundamental to achieving a high personnel readiness.

Beyond personnel readiness is, certainly, material readiness of an individual unit, whether it's an aircraft squadron on a ship, a submarine, or a Marine battalion. In peacetime, material readiness becomes quite critical, and it's the responsibility of all levels of command to ensure that the fleet or force is not driven or operated to such a high degree that material readiness and personnel readiness decline to an unacceptable state. Here we must seek a balance of fleet readiness that is dependent to a large degree on achieving adequate operating time, which includes adequate steaming hours, adequate flying hours, below which the overall readiness of the fleet declines. Obviously if we drive the fleet to the degree that personnel are never in their home port, never have an opportunity to maintain their equipment, never have an opportunity to see their family, then in that situation, and certainly in peacetime, personnel readiness goes down, retention goes down, and overall operational readiness goes down, and that is a situation that is not in the best interest of the United States.

"DAMN EXEC"

The manner in which orders are relayed down the chain of command is another leadership issue. The following sea story, "Damn Exec," by LCdr. Stuart D. Landersman, was originally published in the January 1965 U.S. Naval Institute *Proceedings* and addresses the issue in a most effective manner.

"Damn Exec"

The Norfolk wind was streaking the water of Hampton Roads as Commander Martin K. Speaks, U.S. Navy, Commanding Officer of the USS *Bowens* (DD-891), stepped from his car, slammed the door, and straightened his cap. As he approached the pier head, a sailor stepped from the sentry hut and saluted.

"Good morning, Captain."

"Good morning, Kowalski," answered Commander Speaks. He took pleasure in the fact that he knew the sailor's name. Kowalski was a good sailor.

He had served his entire first cruise in the *Bowens* and did his work well.

The Captain noticed that, over his blues, Kowalski wore a deck force foul weather jacket, faded, frayed, dirty, and spotted with red lead. "Little chilly this morning," said the Captain as he walked by. "Yes sir, sure is," replied the sailor with his usual grin.

As the Captain approached his quarterdeck, there was the usual scurrying of people, and four gongs sounded. "Bowens arriving," spoke the loudspeaker system, and Lieutenant (j.g.) Henry Graven, U.S. Naval Reserve, gunnery officer and the day's command duty officer, came running to the quarterdeck. Salutes and cheerful "Good mornings" were exchanged, and the Captain continued to his cabin.

Lieutenant Graven looked over the quarterdeck and frowned. "Let's get this brightwork polished, chief."

"It's already been done once this morning, sir," replied the OD.

"Well, better do it again. The Exec will have a fit if he sees it this way," said Graven.

"Yes sir," answered the OD.

As soon as Graven had left, the OD turned to his messenger, "Go tell the duty boatswain's mate that Mr. Graven wants the brightwork done over again on the quarterdeck."

Later that morning, Captain Speaks was going over some charts with the ship's executive officer, Lieutenant Commander Steven A. Lassiter, U.S. Navy. The Captain had just finished his coffee and lighted a cigarette. "Steve, I noticed our pier sentry in an odd outfit this morning. He had a foul weather jacket on over his blues; it looked pretty bad."

"Yes sir. Well, it gets cold out there, and these deck force boys have mighty bad-looking jackets," the Exec said.

The Captain felt the Exec had missed his point and said, "Oh, I realize they have to wear a jacket, but for a military watch like that, I'd like to see them wear pea coats when it's cold."

Lieutenant Graven was talking with a third-class boatswain's mate on the fantail when the quarterdeck messenger found him. When told that the executive officer wanted to see him, Graven ended his discussion with, "There, hear that? He probably wants to see me about the brightwork. I don't care how many men it takes to do it, the Exec told me to be sure to get that brightwork polished every morning."

The executive officer indicated a chair to Graven and asked: "How's it going these days?"

Lassiter had always liked Graven, but in the past few months, since he had taken over as senior watch officer, Graven seemed to have more problems than usual.

"Okay, I guess," Graven replied with a forced grin. He knew that things were not as they used to be. It seemed strange, too, because everyone on the ship had been so glad to be rid of the previous senior watch officer, that "damn" Lieutenant Dumphy. The junior officers even had a special little beer bust at the club to celebrate Dumphy's leaving and Graven's "fleeting up" to senior watch officer. Now the Exec was always after him. The junior officers didn't help much either, always complaining about the Exec. Maybe the Exec was taking over as "the heel" now that Dumphy was gone.

"That's good," said the Exec. "Here's a little thing that you might look into. These men who stand pier watches have to wear a jacket, but the foul weather jacket doesn't look good for a military watch. I'd like to see them wear their pea coats when it's cold." Graven had expected something like this, more of the Exec's picking on him. He responded properly, got up, and left.

Graven told his first lieutenant: "The Exec says the pier head sentries can't wear foul weather jackets anymore. If it's cold they can wear pea coats," he added.

"But the pea coats will get dirty, and then what about personnel inspections?" asked the first lieutenant.

"I don't know," Graven shook his head, "but if the Exec wants pea coats, we give him pea coats!"

"Pea coats!" said the chief boatswain's mate, "Who says so?"

"That's what the Exec wants," said the first lieutenant, "so let's give him pea coats."

"The Exec says pea coats for the pier sentries when it's cold," announced the chief to his boatswain's mates.

A third-class boatswain's mate walked away from the group with a buddy, turned and said, "That Damn Exec. First I got to have all my men polish brightwork on the quarterdeck, now they got to wear pea coats on sentry duty 'stead of foul weather jackets!"

Seaman Kowalski's relief showed up at the sentry booth at 1150. "Roast beef today," constituted the relieving ceremony.

"Good, I like roast beef," was the reply. "Hey, how come the pea coat?"

"Damn Exec's idea," said the relief. "We can't wear foul weather gear no more out here, only pea coats."

"Damn Exec," agreed Kowalski. "Captain didn't say nothin' when he came by."

"The Captain's okay, it's just that Damn Exec. He's the guy who fouls up everything," complained the new sentry.

Seaman Kowalski had just gone aboard the ship

when Captain Speaks stepped out on deck to look over his ship. The quarterdeck awning shielded the Captain from the view of those on the quarterdeck, but he could clearly hear the conversation.

"Roast beef today, Ski."

"Yeah, I know, and we wear pea coats from now on."

"Whaddaya mean, pea coats?"

"Yeah, pea coats on the pier, Damn Exec says no more foul weather jackets."

"Well that ain't all, we got to polish this here brightwork 'til it shines every morning before quarters. Damn Exec says that too."

"Damn Exec."

Captain Speaks was shocked. "Why 'Damn Exec' from these seamen?" he thought. It was easy to see that the executive officer had passed the order along in proper military manner. It was easy to see that the junior officers, leading petty officers, and lower petty officers were passing it along saying "The Exec wants. . . ." That's the way orders are passed along. Why? Because "it is easy."

"All ship's officers assemble in the wardroom," the boatswain's mate announced on the loudspeaker system. Lieutenant Commander Lassiter escorted in the Captain. The junior officers took their seats when the Captain was seated. The executive officer remained standing. "Gentlemen, the Captain has a few words to say to us today."

The Captain rose and looked around slowly. "Gentlemen, we are continually exposed to words like administration, leadership, management, capabilities, organization, responsibilities, authority, discipline, and cooperation. You use these words every day. You give lectures to your men and use them, but if I were to ask each of you for a definition of any of these words I would get such a wide variety of answers that an expert couldn't tell what word we were defining. Some we probably couldn't define at all. We still use them, and will continue to use them as they are used in the continually mounting number of articles, instructions, and books we must read.

"If I were to ask any of you how can we improve leadership I would get answers filled with these words—undefined and meaningless.

"If we listed all of the nicely worded theories of leadership, studied them, memorized them, and took a test in them, we would all pass. But this would not improve our ability as leaders one bit. I can tell a story, containing none of these meaningless words, that *will* improve our leadership.

"In 1943, I was secondary battery officer in a cruiser in the South Pacific. In my second battle, gun control was hit and I lost communications with everyone except my 5-inch mounts. I could see that the after main battery turret was badly damaged and two enemy destroyers were closing us from astern. At the time my 5-inch mounts were shooting at airplanes. I ordered my two after 5-inch mounts to use high capacity ammunition and shift targets to the two destroyers closing from astern. 'But Mr. Speaks, we're supposed to handle the air targets; who said to shift targets?' my mount captain asked.

"There were noise and smoke and explosions that day, but the explosion that I heard and felt was not from a shell, but from those words of the mount captain.

"Those attacking destroyers got a few shots in at us before we beat them off. Maybe those shots found a target and some of my shipmates died. I never found out. There was too much other damage.

"I thought over the battle afterward and realized that this entire situation was my fault, not the mount captain's. I may have been responsible for the death of some of my shipmates because up to that day I always gave orders to my subordinates by attaching the originator's name to it.

"What does that mean? It means that it was the easy thing to do, to say, 'The gunnery officer wants us to shift targets.'

"In this peacetime world you may say that we no longer have this struggle on a life or death basis. Quick response does not mean life or death now, but it might tomorrow, or sometime after we've all been transferred elsewhere and this ship is being fought by people we don't know.

"Whether you're cleaning boilers, standing bridge watch, or administering your training program, it's easy to say 'The Exec wants' or 'Mr. Jones says.' It's the easy, lazy way; not the right way. You can sometimes discuss or even argue with an order, but when you give it to a subordinate, make him think it is coming from you.

"Giving orders the lazy way is like a drug. Once you start saying 'The ops officer wants,' you will find yourself doing it more and more until you can't get a thing done any other way. Your men will pass along orders that way, too, and it will become a part of your organization right down to the lowest level. When some problem arises and you want action, you'll get 'Who wants this?' or 'Why should we?'

"Each of you ask yourself if you have given an order today or yesterday in the lazy manner. I think almost all of us have. Now ask yourself if that order really originated with the person who gave it to you, or did he receive it from a higher level? We never really know, do we, but why should we even care?

"In almost every unit the 'lazy' ordering starts on a particular level. From personal experience I can tell you that this can be an exact measure of the unit's effectiveness. If it starts at the department head level or higher it's a relatively bad outfit, and if it starts at the chief's level it's a relatively good outfit. You can find the level below which it starts by hearing a new title preceding a primary billet. 'Damn Exec' means that the executive officer is the lowest level giving orders properly. 'Damn division officer' means that the division officers are taking responsibility for the order.

"Here I am using some of those words, responsibility and authority, those undefined terms we want to avoid, but perhaps we have helped define them.

"To be more specific, every officer does some 'lazy' ordering, but we need to do it less and less. We must try to push the 'damn' title down as far as it will go.

"Let's push the 'damn officer' down all the way to the chiefs and below, then we will have a Damn Good Ship."

MCPON Sanders: *We must be very supportive of seniors. That doesn't mean covering up anything, but if you speak down about your seniors to subordinates, you're not really building yourself up, in fact you're also tearing yourself down. If you have a problem with a senior I suggest you talk it over with your peers and then address the matter up through the chain of command. While that may not always be easy, and sometimes you may not be able to do it at all, never take it up with subordinates, because they cannot do anything about it.*

Chapter 3. Study Guide Questions

1. What is a fundamental assumption every leader must make?
2. What does Admiral Long say about decisions made contrary to one's beliefs?
3. What should one do when faced with difficult decisions?
4. What does Secretary Webb have to say about an officer who adheres to his or her value system?
5. What does Admiral Ramage say about building the strength of a unit?
6. What is the major point made in LCdr. Landersman's story about the "Damn Exec"?
7. What advice does MCPON Sanders give about what to do if you have a problem with a senior?

Vocabulary

unequivocal
enunciate

Chapter 4. Carrying Out Orders

There is little rational argument that can be raised about the role and legitimacy of discipline in a society or its separate parts. It is the stuff and substance that turns an unruly mob into a cohesive group, banded for specific objectives. Discipline is a major advantage to a unit in battle. Discipline takes many forms, and the military record is replete with demonstrations of the obvious fact that the disciplined fighting force, be it an army or an individual, almost invariably carries the day against the undisciplined, even though the latter might be superior in numbers, machinery, or terrain. This section concerns discipline as it relates to obeying orders.

There is no magic formula for achieving promotion and success. It has been established that an individual must learn to be a good follower in order to be a good leader. It follows naturally that the leader should comply with the decisions, the orders, and the directives of seniors and begin assignments as promptly as possible after they are received. Adm. Takaichi Itaya states that

an officer must always fully comply with the decisions, orders, and directives of his seniors. However, this must be on the proviso that these orders and directives are within the authority of the seniors, that they are legally and morally appropriate, and, furthermore, that they do not exceed the capability of the recipient.

An example of a military force that went beyond the bounds of morality is the Nazi SS, who were so self-contained and so driven to carry out their duty that they were ignorant of the harm that their actions caused. When the Nuremburg tribunal charged an *Einsatzgruppe* with averaging 340 killings a day and peaking at 700 per day for as long as 30 days at a time, the German commander, to the horror of those in the courtroom, showed that he was proud of those figures. He told the prosecution that he did his duty as best he could at all times.

Admiral Itaya continues:

An officer must bear in mind that his seniors have one more level of responsibility and have greater information. Regardless of this, if these orders and directives cannot be complied with, the officer will state his view to dispel doubts. When the situation is impossible to resolve, the way is open for him to appeal to a higher level of authority. However, in such a case he should be prepared to accept responsibility for the disruption that this will cause.

Modern business and management courses teach us that leaders are democratic, autocratic (one person rules), or participative (a group or committee acts as a body to give directions) in their approach to solving problems. These approaches form a practical basis for developing an individual method of managing and leading. However, it seems that both leaders and subordinates forget at times that they live and function as military members in an autocratic system by personal choice. All of them raised their right hand and swore, among other things, to obey the orders of those appointed over them. This oath commits them to meet obligations and fulfill responsibilities not encountered by those outside the military. Admiral Long emphasizes this point by stating that

> command involves responsibility, and responsibility is something that an individual either has or does not have. People who are in the chain of command must clearly recognize that they can delegate most or all of their authority, but they can never divest themselves of their responsibility. The commanding officer is 100 percent responsible for what goes on in a missile division; clearly the gunnery or weapons officer is 100 percent responsible; and the division officer is also 100 percent responsible. An understanding and appreciation that responsibility is absolute is critical to an officer doing his job.

> If an officer's boss gives him an order that is clearly unpopular with the troops, he should nevertheless pass that order on as his own. Junior officers must recognize that their loyalty to the commanding officer of their ship, squadron, or battalion requires that if they disagree with a commanding officer's order, after they have had an opportunity to express their disagreement, they take the order and carry it out as their own, whether their disagreement altered the order or not. The military system has worked successfully that way for a long time, and it should continue to work that way in the future.

When a senior gives an order, the junior has an obligation to act promptly and to make the best possible effort to carry it out. If a live steam line breaks and steam endangers those around it, the rudder fails, or they are told to take that hill, military personnel do not want a vote taken on what their actions should be. Lawful orders, no matter how trivial, should be promptly obeyed. Every order has a significance to the overall effectiveness of the ship.

Combat-sustained damage, fire, flooding, collision, and other material- and life-threatening hazards are constant companions and will ever remain so aboard ships of war. Only well-prepared and carefully trained crews can assure survival in times of emergency. Instant obedience is a vital force in their conduct, as anyone who has been to sea can vouchsafe.

When there is no emergency, the same idea applies during day-to-day operations. Seniors expect prompt action to the best of their juniors' ability. "Nothing lowers the effectiveness of a unit faster or further than acceptance of disobedience or deliberate poor performance." At the same time that juniors are being judged and evaluated by their seniors, their subordinates are watching to see how they respond. If junior officers procrastinate or grouse, how can they expect their subordinates to behave any differently when the juniors direct subordinates to accomplish a task? Adm. Jeremy Taylor describes how officers can best decide how quickly they should begin to implement a command.

> The decision is a function of time: How much time do I have before the problem is in extremis? If it is a sudden red light in a cockpit, you've got to do something pretty fast. If it is something due next month, and the decision is made, now it's time to press on and get the job done better than the other fellow can get it done, just taking into consideration all the other factors involved there.

> What you're going to do is size up the problem and get a plan of attack. Get yourself some milestones. You want to use the time that's available to take in more input so that the quality of the decision is better—where you have time, you're inevitably going to do better. Using factors of judgment, I think that the successful naval officer is not going to bite his pipe or go to the powder room or do a lot of dilly dallying. He's going to go to work with everything he's got to come up with the fastest possible way, even for those things being done next month, he's going to get a plan. Sure, it's a plan from which you can deviate, but don't wait until next week and procrastinate. Start right now doing things in a quality way, whatever you decide the plan is going to be and how you're going to get there to solve this problem. Just get in the habit of doing things now and keeping your mind open to gather more facts.

Furthermore, when carrying out an order and tasking sailors and marines, an officer must be enthusiastic, forceful, and fully committed and make it seem as though the task is his or her idea. When officers can't make it look like their idea, they must show outwardly complete commitment and loyalty to seniors and emphasize the positive aspects of each task. This will often require imagination and the squelching of disagreement or contrary ideas, but it will promote unit cohesiveness and provide the best example of the loyalty an officer expects subordinates to show. Adm. E. W. Clexton notes: "[Subordinates] will do their jobs the way you want them to, the right way, as long as you train them properly and then continue to insist on quality compliance."

Procrastination is defined as deferring action, prolonging, or postponing. Thomas Carlyle writes, "It is one of the illusions that the present hour is not the critical, decisive hour." Leaders must be decisive and responsive. From a practical standpoint, since they are going to have to do the job anyway, they may as well do it cheerfully, well, and promptly. Doing so can only help their own image and that of their subordinates and, most important, contribute to the accomplishment of the mission and the unit's reputation for efficiency. Officers are not leaders or even good followers if they do not live up to the ideals and traditional values taught during their precommissioning training. General Wilson is a firm believer in complying with the decisions of a senior.

If you have a responsibility in any field as a staff officer or as a commander and you are asked a question, you forcefully and articulately as possible give your position. As a result, if you are able to convince your commander, fine. If you are not and he makes a decision, you should be given one more chance to say, "I hope you realize these things, Sir, which I want to reemphasize," and if at that time he overrules you, it's your responsibility to say, "Aye, aye, Sir, you have my whole-hearted support and cooperation in carrying it out." I'm not talking about morally indefensible things, which are practically nonexistent.

I think it certainly is not the mark of a good leader or the mark of a good student to wait too long to begin an assignment. Sometimes individuals say "Well, that's so far in the future, it won't happen on my watch. I'll be retired when that comes up. Why should I begin it now, even though I think it's a good idea?" I'm reminded of the story of a commander who wanted to have trees planted near the parade ground, and his staff officer, a G-4, came in and said, "Well, General, we'll get around to those trees sometime, but they'll take twenty years to grow," and the commander very wisely said, "In that case, we'd better get started this afternoon." This applies not only to planting trees, but to other things and other good ideas. Long-range planning is so important. What would have happened to our country if our forefathers had said, "Well, I think

this taxation without representation is probably unfair, but who am I to jeopardize my life and my fortune and my sacred honor for such a foolish thing as that?"

Many of us have seen the movie or play "Mr. Roberts." Was he a good naval officer? He was certainly loved and revered by his subordinates! Should he be respected? We could conclude that the answer to both questions is no, since Roberts did not show or build loyalty toward his commanding officer. Nor did he indicate that he was enthusiastic and positive about carrying out his orders. He was not able to hide his disgust and lack of respect for the commanding officer. Lastly, except for his sacrifice for their liberty, he did not make the jobs or lives of his subordinates any easier. Maybe on his ship these things were impossible, but it is difficult to conclude that Jim Roberts even tried.

Admiral Hayward explains that

the Naval Service is a profession of such a wide variety of responsibilities and interests and technologies that there is almost no job that does not have more tasks and challenges in a day than there is time to do them. Generally speaking, if someone does not like his job, there is an underlying problem with the individual, and that needs to be addressed; it usually is not a question of whether the job is a good one or not.

Obviously, some jobs are much more demanding and more enjoyable than others, and an individual attains the better ones by performance. An individual must do well, because his success is based on how well he is accomplishing his job. His seniors will look at how well he performed before determining whether they throw him into the fire or the next tough job or a job that is less demanding. So the first responsibility is performance.

Regardless of what the job is, an officer can make it a challenge in almost every circumstance. There are jobs that do not follow the career objectives that the individual has. He might want to be an engineering officer very strongly, and if he does not get off to postgraduate school and get his degree and get into that segment, he is not meeting his career objective, and therefore other jobs will be less satisfying. That individual may even leave the service because he is pursuing the wrong track there. But, on balance, if someone wants to be a specialist in a given area, the Naval Service tries to get him there if he works hard enough at it. The individual who wants to be a generalist and strive for command positions, as do most naval officers, must perform well in what he is doing and he will end up in that exact position.

An officer should strive to let his detailer know what he wants, perform well enough so that the detailer will be responsive to his needs, and keep working to stay in the track that he has established in his own mind. If he does not like the job, the first thing he should do is examine whether he is doing a good job of it. Next he should see if the cause of discontent is related to personality problems or leadership issues rather than to the quality of the job per se.

It is entirely possible not to like a job because of the kind of individual who is supervising. If the individual finds poor leadership and poor morale within his unit, he may want to join another outfit, one that is on the fast track. Or he can take it as a challenge instead of as an excuse. He can stick his shoulder to the grindstone and turn that situation around. That really is the challenge of the top-notch leader. Someone with tremendous potential will get in there and fix the problem. Even if he is a junior in the system, he can generate a positive vantage and work on the problem areas.

In assessing a job, the individual must understand his own capabilities, look at the assignment with objectivity, and then make the most of the situation. Odds are that he will not have too many bad assignments with this attitude.

Naval leaders must exude confidence and enthusiasm and take seriously the phrase in their commissions which mentions special trust and confidence. They must attend to professional activities like mess nights, dining-ins, and dining-outs, and they must understand the customs, traditions, and history of the Navy and the Marine Corps. Naval leaders do all these things and much more because they could not stand the embarrassment of being thought of as less than professional. However, the statement of Rudolf Hess, the commander at Auschwitz who stated the following at Nuremburg, must be remembered so that our officers do not repeat such mistakes: "We were all so trained to obey orders without even thinking that the thought never occurred to us to disobey. Someone else would have done it anyway. You can be sure it wasn't always a pleasure to see those mountains of corpses and smell the continual burning, but Himmler ordered it and had explained its necessity. I never really gave much thought to whether it was wrong." What we do must be morally right—if you are in doubt, discuss the matter with your senior and further up the chain of command if necessary.

In the same light, a military professional disdains those who procrastinate or do less than they are capable of doing. They demonstrate a good sense of timing, and because they know their own limitations and those of their people, they listen to advice from those above and below but take full responsibility for a decision, good or bad. If the outcome of a decision is bad, they must never pass the buck. When the outcome is good, they must always try to pass praise and commendation on. If they attain command, the reward for loyalty, moral courage, and smart performance, they must count their blessings and then work harder than anyone else in the unit.

MCPON Sanders: *The naval leader should comply with decisions, orders, and directives of seniors, and begin assignments as promptly as possible following receipt of them.*

Chapter 4. Study Guide Questions

1. What does Admiral Long have to say about authority versus responsibility?

2. How does Admiral Taylor describe how officers can decide how quickly to begin to implement a command?

3. Describe the attitude and behavior of a good leader when carrying out an order.

Vocabulary

rational argument
procrastination
milestone

Chapter 5.   Criticism and the Naval Officer

It is not possible for perfect officers to work for perfect leaders and lead perfect followers, because no one is perfect. For this reason, constructive criticism—the ability of one person in a naval hierarchy to improve the performance of another through objective, corrective comment—is a vital element of a leader's work. The ability to deliver criticism smoothly, to receive criticism with equanimity, and to elicit criticism where it would be helpful is a leadership skill that must be cultivated.

Gen. W. H. Rice points out that

many successful naval officers consider seeking responsibility and taking responsibility to be a key of successful leadership. Young officers should not be afraid to take responsibility because they are afraid of making a mistake. Some young officers put off making a decision because they are afraid they will make the wrong decision. But it is far better that they do *something*, even if it is wrong, than do nothing. If their decision is based on the facts as they know them at the time, if they have taken all this information and used their judgment and knowledge to make a decision, then their seniors will almost always back them up 100 percent.

This does not mean that officers should make a decision out of nowhere just to do something. They must address the facts before making a decision. But if they have done so and they still make a wrong decision, the seniors will likely support them. In addition, the officer who makes a mistake will learn from it and is unlikely to make the same mistake again. That is the value of experience. No one expects an ensign or a second lieutenant to have vast experience, so ensigns and second lieutenants will very often make mistakes. But their seniors will back them up, because their seniors recognize that the only way young officers can gain experience is to be allowed to take responsibility and make decisions.

Seniors get to be seniors not because they're nice guys, but because they were selected to the rank they hold through a conscientious program. They have been selected by a senior board of officers who have the experience and the knowledge to choose tactically and technically proficient people. A staff officer's responsibility is to use his judgment and his experience to provide advice to the senior officer. But once the decision is made, the staff officer's position is to do his part to execute it. Seniors will listen to subordinates who are known for having common sense and good judgment, but seniors look at problems from a different perspective. Seniors look at a larger picture than young officers are able to look at, so young officers cannot expect the senior to accept their recommendations every time. When they have the time, seniors should explain their rationale for their decision, for these explanations are part of the education and experience that young officers need to gain.

ACCEPTING CONSTRUCTIVE CRITICISM FROM ABOVE

Everyone in the Naval Service has a boss—bosses, in fact. Bosses can be expected to have more experience and broader responsibilities than those who work for them. Critical comments passed down the chain of command to the individual officer and leader are the mechanism by which leaders both exercise their responsibility and try to help their subordinates past rough spots.

An officer must pay attention to what the boss says. The officer must pay attention, even if the comments are phrased in the nicest way, are barely critical, or are only slightly corrective. A pleasant civility in delivering course corrections is often displayed by senior naval officers (and this approach is worthy of imitation by their juniors), but course corrections they are, nonetheless, and the boss certainly will note and correct a failure to respond to what was intended to be critical comment.

An officer must pay attention, even if criticism from above is a constant diet. Some commanding officers are screamers, some are nitpickers, some occasionally attempt to revisit scenes of former glory by doing a junior's job (a junior will get a lot of help from these types). But that does not change either the junior's relationship with seniors or the responsibility to respond to their comments. An officer should stay tuned to the leader who talks softly, and not tune out the one who talks incessantly.

Finally, officers should not rely solely on a report of fitness for a clear picture of how they are doing and where they need to improve. Far more useful than the fitness report itself is the chat that goes with it. An officer should ask the boss how he or she is doing and what areas should be concentrated on— and the officer should be responsive to the answers.

OFFERING CONSTRUCTIVE CRITICISM TO SUBORDINATES

Setting standards is an integral part of the naval leader's job. Standards of readiness, standards of appearance, standards of training, standards of safety—standards, standards, standards. And having established how things should be, the leader must use courage, forehandedness, and zeal in holding the high line. That means effectively passing critical corrections to subordinates.

Having passed them, an officer must make them stick. It takes tenacity and patience to lead most people. The young officer must understand that there are subordinates who have neither the ability nor the dedication to perform as directed, and this can lead to frustration in attempting to effect change through criticism. One of the characteristics of poor listeners is that they are usually good bluffers. Seldom willing to admit that they haven't listened, they will fake understanding and take a chance—often with disastrous consequences. An officer can impose great pressure on a subordinate, but the young leader must keep in mind that it is the outcome that is important, and so his or her actions should be aimed at achieving the goal, not just exercising authority.

Effectively criticizing junior enlisted personnel is by no means as challenging as delivering criticism to the chief by a young division officer or to the staff NCO by a platoon commander. Chiefs and staff NCOs are easy people to be afraid of, and even if that dimension is not present, a young officer often thinks that criticizing the chief or staff NCO is presumptuous. Nonsense. The chief is pleased to have the help of a young officer. If there is something the chief needs to be told, then the junior should speak up.

In all this, courage, confidence, a strong sense of responsibility, and a keen awareness of the sensitivity of all persons to criticism are absolutely necessary. In short, an officer should set standards, monitor performance, and—through effective leadership, including critical comment—make the right things happen.

Corrective criticism usually means fixing problems on a case-by-case basis. Beyond this, however, a leader is responsible for regularly reviewing the performance of individuals.

Sometimes in the Naval Service a career enlisted person or officer falls short of being the best performer he or she can be. The sailor or marine has talent and seems motivated but does not seem to be able to "get it together." The good leader will try to improve this person's contribution, but it is a frustrating task, often causing leaders to wish they had had a good shot at various sailors or marines in their formative years.

The fact is that many below-par performers continue to be promoted, yet they remain ignorant of the fact that their performance is not (and may never have been) up to snuff. They were not monitored, counseled, and steered properly early in their careers. Their leaders were too busy or too uncaring or too uncourageous to hold up their end of the leadership burden and regularly review performance. *Regularly* means *as necessary*, but, at a minimum, at least monthly.

Performance review may be very challenging for a leader and, frankly, it takes guts to do it right. Some followers may prove to be impossible to help, but most respond to guidance. Many of the prospective "unfixables" will respond to early, fair, and firm counseling. Sound leaders will regularly review the performance of the individuals in their charge and take action to make these individuals as capable and competent as they can be.

ACCEPTING AND OFFERING CONSTRUCTIVE CRITICISM UP THE CHAIN OF COMMAND

Admiral Uchida notes that

in principle, an officer should welcome any suggestion made in the spirit of good will and hon-esty. By so doing, the officer will foster better cooperation and submission, thus enhancing his corps' morale.

There will, however, be occasions when he cannot adopt those suggestions, for various reasons that are difficult to make public. An apologetic attitude harms the officer's dignity and authority, so he should be straightforward at all times.

It is best if others, by observing the officer's daily work, come to believe that he is always the superb decision maker whom everyone can fully trust.

Wise leaders quietly solicit comments from the subordinates whose opinion they value and also regularly sample the organization to obtain knowledge of what the troops are saying and doing and worrying about. A certain openness to criticism from below can be very helpful here; the inputs serve both as valuable corrective comments and as a measurement of how well or how poorly various policies are understood. All too often the word passed from above either never gets to the nonrated personnel or has been significantly altered on the way down the chain of command. General Barrow points out that

when a junior officer speaks forthrightly and confidently, his seniors will almost always consider his suggestions. The junior who is taking part in a discussion and sees that a decision is about to be made can usually find the right moment to say, "May I make a suggestion, Sir?" Then he must convey his confidence in his ideas by speaking effectively.

The officer can also put his suggestions in writing. Seniors are more likely to spend time considering ideas that are expressed in readable and persuasive prose. Even if the senior officer does not implement the junior's idea, he will make a note of how effectively he presented it. In addition, the junior who writes his ideas down has a record of them and thus can pass them on to others.

To make sure these nuggets passed up from the troops are useful, leaders must do two things. First, leaders need to accept that the comments may have a low validity factor. That's okay. If the problem is one of misperception in the ranks rather than less-than-perfect leadership above, the inputs from below can serve as valuable intelligence on how well the message is being transmitted and received. Second, leaders must maintain an even composure as the comments come in, at least to the extent of keeping the conduit open. It takes courage to tell the boss he or she is mistaken; close that conduit once and it is closed forever.

Adm. David McDonald, wanting to emphasize that it is appropriate for an officer to speak up, gives this example.

In later years, but long before I'd become CNO, a flag officer I was working for said to me: "I will probably seldom ask you for your advice, but if I do, I want to know what you are thinking, not what you think I'd like for you to think." Subsequently, as a commander, I was attending a conference of high-ranking officers, one of whom (not the one mentioned above) I was working for. Many views were being expressed on the subject at hand and not all were in agreement. Ultimately the vice admiral who was chairing the meeting asked for my views. I expressed them, rather vociferously and in direct opposition to the previously expressed views of the vice admiral. Following the meeting, the flag officer for whom I was working chided me a bit and said I shouldn't have talked to the vice admiral the way I did. Shortly thereafter I left my job and went out to the Pacific as air officer and later executive officer of the *Essex*. Upon being detached some time later (after I had been selected to captain), I was advised that I was being sent to Ford Island on the staff of the vice admiral to whom I had talked so rudely, and at his specific request.

Young officers should also be prepared to make appropriate critical comments to their seniors. Tactfully, sensitively, but courageously, junior officers in a command relationship owe the senior officer their best judgment, opinion, and advice. Indeed, in certain situations, a principal responsibility for the junior is careful second judgment. Executive officers have this relationship with their commanding officers. Division officers do well to see that their department heads get their best advice, and platoon commanders should do the same for their company commanders. A simple fact in the Naval Service is that "good followers give their best judgment to their leaders—and good leaders listen." However, it must be remembered that there is such a thing as subordinate power and authority, which is the ability to make decisions based on one's own thoughts, which may or may not be in favor of the senior. An excellent example is when Gen. Theodore Roosevelt, Jr., discovered that his troops had been landed on the wrong beach on 6 June 1944, D-Day, World War II. There was little opposition, but the only way they could move inland was by a narrow road cutting through swampy ground. If the enemy attacked while his soldiers were on this road, the result would

be disastrous. On the other hand, if he waited until a faraway superior told him what to do, it might be too late. "We'll start fighting the war right here," Roosevelt told his subordinates. His divisions moved forward, and their success contributed tremendously to the Allied cause. Had Roosevelt been a weaker man, a tremendous opportunity for victory would have been lost.

DESTRUCTIVE CRITICISM

Officers of the Naval Service, even ensigns and second lieutenants, occupy powerful positions. Irresponsible criticism from a unit's officers can destroy unit integrity and the effectiveness of the individuals in the unit. The conscientious young officer will dampen wardroom "bitch sessions" wherein the Navy or the Marine Corps, the command, or the commanding officer are being harshly criticized, and will attempt to stifle a mean putdown of a fellow officer or an enlisted person of the command.

To be useful, criticism must be pointed somewhere, toward some corrective improvement. Constant, idle complaint is a pernicious habit that an occasional wardroom will inflict on itself. Kill it in the crib. Kill also the urge to hammer some poor sailor or marine in public. Matters of safety and primary operational urgency require immediate corrective action by a leader, but most times and most circumstances permit operation of the classic rule "Praise in public, censure in private." Unless it is your purpose to destroy forever the effectiveness of the person you are correcting, regard the rule as sacred.

SUCCESS THROUGH COOPERATION

Admiral Burke describes how he learned that success is most often achieved through cooperation.

I was very fortunate, because I entered the Naval Academy completely unprepared, as I only went two years to high school. When I got into the Naval Academy there were a lot of little things that I should have known that I didn't, so I stood way down in the bottom of the class, and I had a rough time, barely making it. Well, I got through because classmates helped me. They would spend the time on me in teaching me things that I should have known, and everyone helped me, including professors sometimes. I learned that no man can do very much by himself, that a man does a job and 90 percent of what he's credited for doing is done by someone else. It's either done before him or after him.

Now this happened over and over again, everyone knows it's true, but it takes a long time to realize that any job that is done, is done by a group of people, so you've got to be careful not to think that an individual by himself is all that good. A lot of it is, you happen to have a good chief boatswain's mate, you happen to have a good vice chief, you happen to have a well-trained crew that you inherited from someone else, or you happen to have a kid with bright eyes who could see very well and he saw something that no one else could see and reported it.

I was on many boards [after] I retired from the Navy, and one time I objected to a movement of a plant in England of one of the companies I was with. I objected because I didn't know enough about it. Well, I didn't know enough to say no, but I put in my reservations just the same. They went ahead and moved the plant from the outskirts of London up to the east coast of England. They spent a lot of money, they had to build a plant up there. About a year after they got the plant built, people wouldn't move. Since I had objected, they asked if I would mind going over to England and taking a look, so I went over and talked to a few people. In two days I found out what was wrong. What they hadn't considered was that the British people don't move. They liked their job where they had it, and they wanted to live there. Their grandfathers lived there, and they were not going to move up the east coast. So they had to get new people.

This new young fellow who had put this superior plan in was a brilliant man, but I called up and said, "I've made the decision. You've got to fire this young fellow and get him out of here because he can't run things, no matter how good he is technically, he can't run it because he doesn't know people. You've got to call that old guy back who is retired for age and get him to come back." Well, he came back on a contract for two years at the same salary that he had before, but the important thing is that at the end of two years he would retire. If he had made a success, he would get a big sum of money, I think a half-million dollars or something like that. If he didn't make a success, he didn't get anything more than his pay. The company jumped at that, and of course he made a success. This whole thing hinged on a very minor thing: they had forgotten that they weren't dealing with Americans in the moving thing.

MCPON Sanders: *It is recognized that it may be difficult for a junior officer not to show resentment when a senior enlisted is making a suggestion about a plan that the officer had put together. While they may not speak from the same educational view as the officer, senior enlisted do speak from experience, and sometimes the experience is a little bit better than the booklearning the officer has received. While you don't have to follow the suggestions, do listen to them, and at least recognize that the suggestions and comments were given in good faith, and weren't made to make fun or put down your thoughts and ideas.*

Chapter 5. Study Guide Questions

1. What does General Rice consider to be a key of successful leadership?

2. What is the mechanism by which leaders exercise their responsibility to help subordinates past rough spots?

3. What concerning standards is an integral part of the naval leader's job?

4. How often should a senior review subordinates' performance with them?

5. According to General Barrow, what phrase can be used to good effect by juniors who wish to make their views on an issue known to their senior?

6. What is a characteristic of good followers and leaders in the Naval Service regarding advice?

7. What rule should always be followed regarding the delivery of criticism?

8. To be useful, what characteristic must criticism have?

9. According to MCPON Sanders, what attitude should junior officers take toward suggestions given by subordinate enlisted personnel?

Vocabulary

constructive criticism
equanimity
performance review
pernicious habit

Chapter 6. Moral Responsibility and the Naval Officer

God grant that men of principle shall be our principal men.

THOMAS JEFFERSON

A naval officer is morally responsible for all aspects of what happens to his or her unit. The basis of this principle of naval leadership is codified in Article 0702.1 of Navy Regulations, 1973, which provides that the responsibility of the commanding officer for his or her command is absolute, except when and to the extent, relieved therefrom by competent authority. This rule of absolute responsibility, and its corollaries of authority and accountability, have been the foundation of operational efficiency and effectiveness throughout the history of the Naval Service.

Today's officers must also be aware of the moral, financial, and cultural dilemmas faced by their personnel. They are morally responsible for their welfare. All commanding officers and others in authority in the Naval Service are required to show in themselves a good example of virtue, honor, patriotism, and subordination; to be vigilant in inspecting the conduct of all persons who are placed under their command; to guard against and suppress all dissolute and immoral practices, and to correct, according to the laws and regulations of the Navy, all persons who are guilty of them; and to take all necessary and proper measures, under the laws, regulations, and customs of the Naval Service, to promote and safeguard the morale, the physical well-being, and the general welfare of the officers and enlisted persons under their command or charge.

According to Gen. Bernard Montgomery, "the four cardinal virtues" provide a good guide for moral responsibility and leadership:

1. *Prudence.* Refer all matters to divine guidance. Outcomes are impartiality, wisdom, and tact.
2. *Justice.* Give everyone his or her due. Outcomes are religion, obedience, gratitude, integrity, and goodwill.
3. *Temperance.* Pursue self-control for the highest development of human nature and for personal and social needs. Outcomes are purity, humility, and patience.
4. *Fortitude.* Follow the spirit that resists, endures, and triumphs over temptations of life. Outcomes are courage, industry, and self-discipline.

Absolute responsibility is imposed upon a commander by positive force of law in the form of a written regulation, but the scope of command responsibility is based on more than just codified legal principles. A commission as a naval officer should not be viewed as simply a contract of employment between a person and the U.S. government that creates only legal obligations on the part of the individual in return for monetary compensation. More importantly, the oath of acceptance of a commission as a naval officer must be viewed as a pledge to contribute to the common good of our society as a way of life, making the commitment both legal and, to a much greater extent, moral. The rewards for moral commitment are not necessarily tangible. Officers of the Navy and the Marine Corps, having traditionally sought positions of high trust and confi-

dence in the service of their country, often find that the greatest reward is the personal satisfaction of command. Along with the authority and perquisites of command go the responsibilities of leadership.

CONSCIENCE

Responsibility has been defined as the state of being answerable for an obligation, fulfillment of which is dependent upon an individual's judgment, skill, ability, and capacity. While legal responsibilities are imposed by law or contract, moral responsibilities are recognized and enforceable by an individual's conscience. The conscience can be thought of as the ethical layer of the human personality, which provides people with inherent values about the differences between right and wrong. Like other facets of human character, the conscience is influenced and developed by many sources: family, religion, education, standards of the community, and the ethics of the Naval Service and the nation. An individual assumes moral responsibilities because of the compulsion of conscience, not because of possible legal retribution. For example, you may see a person drowning in a lake. Although you may have no legal responsibility or duty to save that person, you have a moral obligation to render assistance, because the society in which you live has placed a high value on human life. In attempting to help such a person you have assumed a responsibility demanded by society, whereas failing to act would be a refusal to accept such responsibility.

Service as a naval officer is a noble profession that demands many virtues: unselfishness, a sense of duty, honesty, courage, obedience, loyalty, and integrity. While society demands many of these attributes from all people, every one of them is critical in a military context, for, without them, a unit could not operate. An officer who is driven by these underlying values cannot help but do the "right" thing, despite the fact that another course of action might be easier or less dangerous to career aspirations.

It is often necessary for naval officers to address moral considerations in the conduct of their duties. In particular, they must always be concerned about the moral welfare of their personnel. If naval officers were to adopt solely a contractual view of their responsibilities as commanders, they might find it relatively easy to attempt to divorce their military functions from moral considerations. War could be thought of and conducted as a business. Since the primary tasks facing a military leader are the development of strategy, tactics, and weapons systems that will efficiently destroy real and potential enemies, the body count becomes the bottom line. The theory is similar to the adoption of a contractual view in the teaching profession. This concept sees the role of the teacher as a transmitter of purely objective knowledge, packaged and distributed; here, the grade point average is the bottom line. Neither approach accepts responsibility for forming the character of the people being led, and there is no predicting the uses to which their weapons or knowledge may be employed. This is a chilling thought in an era of genetic research and chemical weapons. Leadership is not an impersonal, value-free operation. Approaches that ignore the critical moral aspect of leadership must be viewed as incomplete.

Moral Courage

Failure to act in our example involving the drowning person, although not punishable criminally, may indicate the lack of an element that is inherent in moral commitment and assumption of responsibility. This element is moral courage. Without moral courage (as distinguished from physical courage), the officer's ability to lead others is impaired. Officers must possess moral standards in both thought and action, which includes speaking out when they believe they are right. Firmly adhering to a code of moral values is the basis for integrity, which is essential to strong leadership.

Training for the armed services places an emphasis on readiness to accept responsibility, because leaders must learn the meaning of moral courage early in their careers. If officers fail to learn this when they are young, they will not be prepared for the heavy responsibilities of high command, where great moral courage is demanded in a variety of situations. One moral situation that a commander may be confronted with was described by Vice Adm. William Mack in this retirement speech, delivered upon completion of his duty as the 47th Superintendent of the U.S. Naval Academy.

As I complete 42 years of service, I would like to leave to the Brigade of Midshipmen a legacy of one idea which represents the distillation of that experience: The one concept which dominates my mind is that of the necessity of listening to and protecting the existence of the "Dissenter"—the person who does not necessarily agree with his commander, or with popularly held opinion, or with *you*. Unfortunately, history is full of examples: then-Commander Mahan, whose novel ideas of seapower fell on barren ground; then-Commander Sims, whose revolutionary (but correct) ideas on naval gunnery ran counter to those of his seniors; then-Commander Rickover, who fought a lonely battle for nuclear power. All eventually succeeded, but *not* with the help of patient, understanding officers. Regretfully, each

needed help from outside the uniformed Navy. We cannot afford this way of life in the government or in the Navy in the future, for the intervals given us for discussion and decision will be increasingly shorter. You may ask: What can you, as a midshipman or junior officer, do about this? My answer is that time will go by for you, as a busy naval officer, very rapidly. Before you realize it, in a decade *you* will be the young commander called upon to give your honest (and, perhaps, dissenting) opinion. In another few years, you will be the senior officer charged with preserving and using the dissenting opinion of another.

The point is: To begin at this early age to cultivate an open mind, to determine to hear all arguments and opinions, no matter how extreme they may seem; and, above all, to preserve and protect those who voice them.

In the daily life of the junior officer, small acts of moral courage are constantly demanded in the administration of justice, in making decisions, which, though known to be right, will probably prove unpopular, and in the acceptance of responsibility, particularly when things have gone wrong. Every time an officer turns a blind eye to a situation or behavior that he or she knows to be wrong, such as a minor breach of discipline, the officer is in fact refusing to accept the responsibility of office and is exhibiting a lack of moral courage. A succession of even minor failures may strike at the roots of an officer's personal integrity, and destroy his or her ability to lead.

An example of a succession of moral failures that led to disaster is a West Point cheating scandal that occurred in the mid-1970s. One hundred and fifty-two cadets resigned or were dismissed after plagiarizing on a take-home exam. An internal investigation found that entire companies had gone "cool on honor," in part because they felt that the academy's honor code had become trivialized by being used as a tool to enforce petty regulations. Instead of working within the system to effect changes in regulations that they believed to be unfair or unneeded, the cadets took what appeared to be the easy way out: they disregarded rules they did not like. Classmates tacitly agreed among themselves to ignore the honor code, and the resultant spiral of disobedience led to disaster.

Integrity no doubt comes more easily to some people, and some are more prone than others to live up to the standards they know to be right. Yet integrity is vital to the leader, and the lack of it will be detected quickly in a person in a position of leadership. Naval officers should never forget that they often command young and impressionable people who, even if they are older than or close to the same age as their officers, look to them for guidance. Every act of moral courage is in fact a victory over oneself and provides an example for others to emulate. On the other hand, to give way to one's innate fears or selfishness is to place oneself in danger of developing moral cowardice. Officers who are unwilling to make the effort themselves have no right to demand it of others.

With respect to the officer being a member of a profession that at times may be required to prevail over an enemy, Admiral Zumwalt believes that

> the mission of the armed services is best described as being so ready to kill that they won't have to kill, to be so ready to win that they won't be attacked. It seems to me, after all, that the fundamental responsibility of a democratic nation is to have armed forces of sufficient strength that those, in tandem with an enlightened foreign policy, can produce peace on the terms that are reasonable for our democratic way of life.

Admiral Holloway says that

> an officer is responsible in a moral sense for what happens to his unit insofar as he has control. As an example, assume a platoon is in a blocking action to permit the withdrawal of the battalion. The platoon could very well take heavy casualties under such a circumstance; in fact, the leader could lose the entire unit. The leader cannot be held morally responsible for the losses unless through a lack of judgment or professional skill he failed to employ his platoon in the best possible way to carry out the mission, or unless he attempted, after making contact with the enemy, to withdraw. Those are the only circumstances under which he could be held responsible, because he had control of that, but the fact that he was put in a blocking position, was left behind as the rear guard, was beyond his control. The moral responsibilities of that could go all the way to the President who declared the war in which he was fighting.

Moral responsibility is difficult to define and regulate because it depends on what a person is made of. No amount of thought, desire, education, or experience will make an officer fit to assume command and exercise moral leadership unless they are accompanied by a conscious, unremitting effort at self-control and self-improvement. Moral responsibility is active, not passive. It is more than just not doing bad things. It is an ongoing process of personal struggle against weakness, expedience, and self-

doubt, and its rewards are strength of character, resolution, and confidence. Admiral Hunt believes that

> there is a very clear need for a military officer to have an understanding of fundamental principles. Perhaps I could give an example. It may be in certain types of warfare that there is a temptation for an officer to use force on some other person, for example, torture, which would normally be of horror to him. For example, if somebody is picked up in the street and you know that he has vital information, there is very often strong motivation, especially in the heat of warfare and battle, to get that information out of him by one means or another or to tell somebody else to make sure he gets the answer. Unless you think about these problems in advance, the chances are that that's what you'll do. So I think that a military officer must have a very clear idea of what is right and wrong before he's faced with these problems, and then when he comes to these problems he doesn't step down the wrong line. I would never ever allow anyone in my organization to break the moral code in that sense, I would never give instructions to do that, and I'd never do it myself—at least that is my view. When faced with a particular problem I can understand that, in the heat of the moment, people might look in a different direction, but I think that would be absolutely wrong.

ACCEPTING MORAL RESPONSIBILITY

The commander's responsibility for his command is absolute. Admiral Nakamura observes that an officer is morally responsible for what happens to his unit. He must be held accountable for its safety, well-being, and efficiency, both legally and morally. Every day tests the strength of character, judgment, and professional abilities of those in command. In some cases, commanders will be called upon to answer for their legal responsibilities in a court of law. Virtually every day, however, officers in the Naval Service will be answerable for their moral responsibilities to their unit. A naval officer should want it no other way, for the richest reward of command is the satisfaction of having measured up to these highest of standards.

In view of the experiences of civilization just before and during World War II, the words of Admiral Clausen are particularly meaningful:

> Responsibility, in a very short explanation, is the necessity or the obligation to give an answer for one's own activities or those of others. In order to undertake responsibility, a naval officer needs hu-

man and technical qualifications and competence in the task conferred on him, which include knowledge and experience. For a commanding officer, the responsibility is comprehensive, indivisible, not transferable, and continuous.

In a difficult situation or during a war, the holder of responsibility comes very often into conflicts, and he has to decide between different duties. To deal with all aspects there are powers given to the holder of responsibility written in the law and in instructions, orders, regulations, etc. By this means he can carry out his task of leadership.

Although this is certainly correct, I do have some additional views from my own experiences.

1. I do not think that the holder of responsibility should cling only to regulations and paper work. Responsibility must go into very abstract places: into the human conscience, into the character, into the mentality, into the holder's power of comprehension. [Editor's note: Traits that lessen an officer's ability properly to care for the welfare of his or her personnel are self-interest and extremes in professional dedication, which lead to "mindless commitment."]

2. Responsibility needs mutual confidence between the leader and the subordinates. This correlation of authority and confidence is especially indispensable during a war or during a difficult situation.

3. Subordinates are not human beings who can be programmed and disposed of in an automatic way. Subordinates have a body *and* soul. That implies that each subordinate who has to carry out a task must be guided to think or, better, to think under consideration of all necessary contingencies, so that he is in genuine connection with his leading officer and the task.

These three additional views make it clear that naval officers must be able to cope with a reality that is never as chemically clean as simulated by theory. Therefore, an inadequately or overly intellectual naval officer will be hindered when he has to act as leader of personnel on board ship. In the profession of a naval officer, very many circumstances (especially by development of technique) have changed and shall change in the future, but the qualifications that are required of the leader will not differ from those that have been valid for centuries.

HONOR

Admiral Clausen discusses honor in the context of moral responsibility:

The word *honor* became, after the war in Germany, in connection with the profession of an officer, discredited. To be outspoken, I never understood exactly what people meant. Obviously they did not want officers to have or own a "special honor" in the same way that they did not like to have the maintenance or cultivation of the idea of an "elite." Now, I think it is not as simple as the ideas behind these words as treated by the public, because these ideas make an enormously high demand that most members of the public would not be willing to meet. A wrong sense of honor is, of course, detrimental to the officer's profession and would be disastrous.

As the honor of profession imposes the highest commitment/liability/obligation on an officer, and furthermore his profession requires him to observe a number of virtues, it seems to me no surprise that a naval officer needs in a well-meant sense a conception of honor that is based on the interest of his profession. A naval officer has a very special job that can't be compared, for instance, with a baker's job. That does not mean that a baker should be respected less.

As the Naval Service becomes increasingly more dependent on technology, its officers may well be increasingly tempted to judge success in terms of systems efficiency—the managerial position—rather than in terms of personal values. No doubt these criteria are vitally important within the context of any complex organization, but the danger arises when the system is emphasized at the expense of the men and women who make it work.

Admiral Clausen continues:

To serve the country must be an honor. I cannot imagine an officer without this interpretation. An officer will be entitled to this concept of honor only if he is educated accordingly and is willing to follow the virtues prescribed to his profession: to serve faithfully; to exercise loyalty, uprightness, veracity, bravery, sense of duty, discretion, comradeship; to perform carefully all duties, especially where subordinates are concerned, etc. These virtues can be learned in theory. That is not of sufficient use. The concrete application of these virtues in practice is of decisive significance. Therefore, an officer must be observed during his career carefully and criticized in pertinent reports with regard to his character, his personality, and his conduct. When this is done seriously, then the result will be expressed in bearing and reputation of the whole community of officers, and I am convinced there will be no difficulty in attaching the word *honor* to the right place.

MCPON Sanders: *The officer is morally responsible for all aspects of everything that happens to his command.*

Chapter 6. Study Guide Questions

1. What are the four cardinal virtues of moral responsibility and leadership, according to General Montgomery?
2. What is responsibility?
3. According to Admiral Mack, what should be a paramount goal of a junior officer?

4. What happens when a leader turns a blind eye to a situation or behavior that he or she knows to be wrong?

Vocabulary

dissolute practices
prudence
justice

temperance
fortitude
perquisite of command

Chapter 7.   Responsibility and the Naval Officer

One of the differences between leaders and those who follow is the propensity on the part of the leaders to seek out situations in which they can contribute and take charge as necessary, as well as their willingness to accept responsibility if something goes wrong. In addition to seeking out opportunities in which their experience can be brought to bear in directing the actions of others, officers are expected immediately to accept responsibility for carrying out any assignment made by their seniors. To Admiral Itaya, responsibility is

our most important asset. A man without responsibility does not have the character to be a member of the organization. Such a person is a detriment to the unit and of no value. During a shelling in World War II, what kept me going was my sense of responsibility. One could say that responsibility was my spiritual savior. One of my superiors told me, "It is not necessary to be ashamed of being afraid. But, be ashamed of acts of military cowardice. One can do nothing more than devote one's self to fulfilling his responsibility."

Admiral de Cazanove believes that

an officer shows he is responsible by, first, searching for responsibilities. The worst thing for an officer to do is spend his time in the service without seeking out additional responsibilities. If you are going to be an officer, then you should work at being a leader, and sneaking around trying not to be in trouble is not the way to be a leader. In addition, you must develop this desire for increased responsibility in your subordinates. The more responsibility you give to others, the better it will be for the job, and you will grow by the

very act of thinking about and assigning responsibilities to others. In other words, you will increase your own capabilities and ability to take on additional responsibilities by learning how to maximize the use of everyone in the organization. Of course, you must always accept the responsibility for everything that takes place in your unit, and, when passing out directions, you should take the responsibility for giving the directions and not say that the troops have to do something because someone else made the decision that something had to be done. You must act as if it was your own personal idea and not the idea of the boss. This is an intellectual necessity.

The size of armed forces makes it impossible for any one individual, no matter how talented, to wage and win a battle alone, and thus it falls on the team, made up of subordinates and their leaders, to put forth a maximum effort to prevail in battle. Of all the responsibilities of a junior officer, mission accomplishment is and must always be the highest priority. While individuals are not born with the desire to excel and to meet the requirements handed down by their seniors, this desire can be learned, and it is the responsibility of the senior to train and educate subordinates in such a way that they will want to do all that is requested of them as well as take initiative when and where necessary to ensure that the overall objectives laid down by their leader are met.

General Rice describes the best attitude for an officer to develop regarding responsibilities.

As an individual advances in any organization, his risks get higher, partly because his subordinates increase in number. An officer gets paid for taking responsibility, and the higher in rank he goes, the more responsibility he will have. He should never shirk from taking responsibility, even though in every officer's career there may be some very unpleasant duties.

This is not to say an officer is expected to volunteer for unpleasant duty, but if he is given an unpleasant assignment, he will, if he is a true professional, get on with it. He will not moan or complain about what he has to do; rather, he will do it better than anyone ever expected it to be done. This is what marks that officer as a true professional, and marks him by his seniors as an individual who has a future, who will not shirk responsibility. This officer will have self-satisfaction, and very likely his senior officers will congratulate him on the job he did. This, of course, is a hallmark of good leadership: telling people they have done a good job as freely as telling them they have done something wrong.

VOLUNTEERING

While only a few people learn in school to volunteer information and their services, officers are expected to keep themselves apprised of the needs of their seniors, subordinates, and peers and to offer to help them in any way. Those in the Naval Service are a part of a team in which no member or unit can consider itself successful unless the next higher or adjacent member or unit is also enjoying success. One of Gen. George Patton's fundamental principles of leadership was that your men always serve with you, not under you. Officers are evaluated by the enthusiasm they show to seniors in making themselves a part of the overall organization. According to General Patton, the most important thing to keep in mind in picking the right leader is to base the decision on ability, not friendship. "School days" may have put the stigma of "apple polishing" on anyone who appeared to show an above-average interest in studies, but officers will soon find that when they diplomatically offer to help, and keep their offer low key, seniors will gratefully accept the offered aid.

Adm. Thomas Moorer observes that

in the Navy, the ultimate assignment is one of command, whether of a fifty-foot minesweeper or a ninety-thousand-ton aircraft carrier. An officer who is seeking command indicates by his actions that he is willing and eager to accept responsibility. Once he takes command, of course, there is no question about who is responsible.

In a staff situation, responsibility for, say, an operation is more diffuse, but in the final analysis the commander is responsible. The commander, in turn, of course, is watching all the people who are working for him on the same basic problem.

In the Naval Service there are many opportunities to seek and acquire responsibility, and the young person who is eager and willing to accept responsibility will find that his service reputation reflects his attitude. During the war, for example, certain pilots would volunteer for night raids because they liked the idea of being in charge, even though it may have been of only two or three aircraft. And these pilots developed a fine reputation.

Officers should constantly try to improve their professional knowledge and capability, and thus officers are rewarded for volunteered efforts with an increase in knowledge, which in turn means that others will select them for assignments that require the knowledge and understanding they have accumulated. The additional understanding gained by junior officers through volunteered efforts is likely to be rewarded by even more jobs. Thus, enthusiastic professional officers who seek out responsibility will find that they are just the kind of person their seniors like to have around, and that when "good" assignments and opportunities to represent the command come up, they will be selected, because the seniors know that such an individual's actions will reflect favorably, not only on the junior officer, but also on the senior who made the selection.

THE DYNAMIC OFFICER

General Barrow states that

everyone has known someone who can be a good leader for a specific occasion, to meet a specific requirement, or at a specific time, but before the event started and when the event is over, he is asleep on the bench. That is his style: He does a great job when he is called on to do it, but only then. This individual has narrowed his capability.

An officer who is a complete officer—not one who sleeps most of the time while waiting for a call to duty—is dynamic. He seeks out responsibility by using his imagination to find things to do beyond the requirements of his assignment. Very few orders limit the officer's search for responsibility by saying "Do this and nothing more."

General Barrow describes a dynamic officer:

Don't be willing to say, "Well, I've read all the orders, and it looks like all I have to do is make sure the place is opened up at eight o'clock and everything is secure and no classified material is on the desk and all the parking lots have been cleared out at seventeen hundred." If you want to be that kind of person, fine, but you won't get anywhere. I'm surprised you even got in in the first place. I think that whatever you've been told your job is, whatever you know it to have been historically, traditionally, or whatever your orders say, implicit in that is one enormous array of unstated things that you can seek out to do. Some of them will come to you by suggestions from subordinates moving about and listening.

There is just no end to things a dynamic officer wants to do. This often draws a distinction between a mediocre officer and an outstanding one. The compliments that the dynamic officer often gets paid by a senior are along the lines of: "Boy, he's something else. That guy not only does everything he's supposed to do, but he can always find some things that need to be done and does them so well." The only limit is you cannot intrude on others or overburden your subordinates to the extent that all you're doing is making yourself look good at their expense.

If you are comfortable in being a commander, if you are comfortable in being a leader, you should be willing to bare your soul. Say "I believe in this, this, and this, and I don't believe in that." As regimental commander, you gather the staff and maybe the battalion commanders, if they are around, and often at night, when you are involved in combat in the field, you just talk and you get messages across, such as the necessity to seek out responsibility. You convey those kinds of messages which you would not put in an order. You wouldn't say, "I expect every man in this command to seek out responsibilities," but you can encourage it too by speaking to the subject, and you set the example sometimes. They say, "The colonel seems to think things need to be done, and nobody else did them, so maybe I ought to go look for some things that need to be done." You reward those who do with praise, which encourages others to say, "The colonel paid Sam a compliment about doing this thing that needed to be done. Maybe I ought to go look for some of those things."

An officer can quite easily avoid additional assignments: by not being around much, by looking at the floor when a senior is asking for help, by making "a big deal" of all the work he or she has to do. Leaders rarely have time to cajole people to take assignments, and they find it discouraging to ask someone to do something and have that person turn them down. So the officer who wants to establish the reputation of not being available and of not wanting to do extra work will have no trouble.

Good officers do not want to make mistakes, and when someone helps them learn from a mistake, they are grateful. In the long run, they will feel pride in themselves for compensating for an error.

As commander of a destroyer, Adm. Robert Carney helped one young officer learn responsibility.

I required each of my officers to maintain a journal. I had one young officer brighter than hell and lazier than hell, and he got way behind on his work, so I told him he was restricted on board until it was up to date, and he said to me that he would have his journal ready by Saturday. I said, "Unfortunately, that won't be useful as far as you're concerned because I won't be here Saturday, so you can have the weekend aboard ship." So I came back Monday, and I was in the Army-Navy Club talking to the clerk at the desk. There was a gentleman standing beside me, he turned around to me and he said, "Are you the skipper of the *Reed*?" I said, "Well, yes, Sir, I am." He said, "Then you're the one who put my son under hack for three weeks." Well, I thought, what's this old guy trying to do, trying to tell me how to run my ship? I got the hair up on the back of my neck right away and I said, "Yep. He didn't measure up to my requirements and I wouldn't turn him loose until he did." He said, "Well, thank God somebody took hold of that brat."

In a basketball game, suppose one player always hangs back near his own basket. His fellow players will exclude him from most actions because they have learned he is not dependable, and the coach will soon thereafter remove him from the game. A reference to sports with respect to officers doing their duty is made several times in this volume, and this is one appropriate point to remark that officers are team players who not only seek out responsibility, but also eagerly accept responsibility when it is offered by their seniors, or even when they are asked to do so by peers and subordinates. Accepting responsibility is another way officers can earn psychic income, the amount of which is limited only by the desire of and the time available to the officer who looks for ways to contribute to the organization.

DEVELOPING A SENSE OF RESPONSIBILITY IN SUBORDINATES

An officer can develop a sense of responsibility among subordinates in a variety of ways, but the most effective method is for the officer to demonstrate by example that seeking out and accepting responsibility is the way to succeed. For example, when officers receive an assignment from their seniors, they can show pleasure in the assignment to the troops; they, then, will know the proper attitude to take when they are given a chance to contribute. Officers who act enthused and quickly get started on their work set a tone that will filter down quickly to the rest of the organization. Officers who volunteer to help with the overall mission also send a message to subordinates that the right way to play the game is "One for all and all for one." While it is certainly not necessary to reward every voluntary action with a three-day pass, a few words or a short note to the individual who has excelled in an assignment beyond normal duties will build a sense of loyalty in the individual which will last for a very long time. General Rice believes that

> it is bad practice for a junior officer to dodge responsibility. There are opportunities for a young officer to take responsibility, and he should not wait for someone to ask for a volunteer, he should step forward. The young officer who sees a need and decides to take charge, to say, "I'll do that," and does not wait to be told, will learn the value of looking for ways to get involved.

An officer gives his subordinates a sense of responsibility by giving responsibility to them. That is a difficult thing for a senior to do, because he has to sit back and give someone the assignment knowing full well that he could do it faster and better himself. But the officer who continually does things himself is doing a disservice to the people under him. So, even though it may be very difficult to sit back and let someone else do something, that is the only way subordinates can be taught to make decisions and to take responsibility.

Once he has assigned to subordinates the responsibility for a task, the officer must hold them to it and make them responsible. He cannot slack off. If he expects something to be done, and he has given subordinates the responsibility to do it, he must hold them responsible. He should not say, "Well, they're young, and I should not have expected that much." If they make a mistake, the senior should let that mistake teach them. He should let them know what their mistake is.

RESPONSIBILITY CANNOT BE DELEGATED

While officers can delegate to subordinates specific authority to accomplish an action in their absence, it is important to remember that responsibility cannot be *delegated*, although responsibility for a mission may be *assigned* to someone else. In other words, although someone else may be given responsibility to see that a certain task is undertaken and completed, the responsibility for the success of that action still lies with the senior officer involved. After Gettysburg, Gen. Robert E. Lee sent a note to Maj. Gen. Cadmus M. Wilcox, making this point: "Never mind, General, all this has been MY fault; it is I that have lost this fight, and you must help me out of it in the best way you can."

Thus, division officers are responsible for everything that happens in their unit. They cannot make the excuse to department heads that the order was passed down correctly but the troops failed to get the job done correctly; this does not relieve officers of the responsibility for proper completion of the task. In the first case, making such a statement will quickly lose the officer the loyalty of subordinates, who have done their best to succeed. Second, it will make an officer look weak in the eyes of seniors, for the officer will be indicating an unwillingness to accept responsibility. All department heads were at one time division officers, so they are fully aware that, no matter how good the officer is, he or she is not capable of doing, and should not do, everything within the unit, and that the officer must rely on subordinates, not all of whom the officer chose. So, while department heads know that officers are responsible for what goes right or wrong, they will usually take the factors involved in an unavoidable failure into account, making it unnecessary for the junior officer to ask for special consideration.

ACCEPTING THE RESPONSIBILITY FOR MISTAKES

If an officer has taken every step possible to train, educate, and inform the unit of its duties and of how it is to meet unexpected emergencies, and he or she follows up with constant practice and opportunities for subordinates to improve their performance, a failure will not end the officer's career, but will be seen for what it is: an unavoidable happenstance. Officers must remember that it is critically important for responsibility to be accurately placed in the service. Only in this way can seniors determine what areas require further attention, so a disaster can be avoided in the future. The work responsibilities of officers are certainly extensive, and time does not usually allow for protracted investigations into what went wrong and why. Thus, officers are expected immediately to volunteer information concerning events with which they were involved, knowing full well that the Naval Service's mission is inherently dangerous and subject to errors due to enemy action and the minute details that must be handled, even during peacetime operations. Officers should do the best they can, always look for methods to improve their performance, take responsibility for their own actions and those of their unit, and above all, seek out every opportunity to volunteer their services to do what is best for the unit, the Naval Service, and the country.

Admiral Clausen offers the following in this regard.

> Although incidents during a war normally produce lessons preponderant in tactical, technical,

operational, and other details, there is one incident that made me more pensive in a general way. This incident happened when I was sunk in 1943 in the English Channel as commanding officer of one of our fleetsweepers (crew of 110) after an engagement with *Hunt*-class destroyers, and I lost half my crew, either dead or wounded. This incident gave rise to a great number of lessons learned, which led me to evaluate, to the benefit of my activities, my time in the Bundesmarine. Out of these more general lessons learned, I shall point out those which might be of interest in this connection.

1. Leadership on board is not the business of one person (the commanding officer) only, but a task of teamwork of all kinds of superiors on board a ship in their different fields and roles of activity. The commanding officer, as holder of the responsibility, is in possession of the key to the ignition and of the control of this process in order to direct and to control the manner of how the crew has to be led. He must find the way to transfer his intentions, imaginations, and ideas of leadership to his officers, chief petty officers, and petty officers and, via this chain, to the ratings.

2. The significance of all petty officers in the field of leadership, training, and education is eminent on account of their permanent and direct contact with ratings. The former Kriegsmarine has learned in this regard very much after the revolution in 1918. The proximity of contact between officers and men became more and more narrow, perhaps even enhanced during the Nazi-time, because no group of society is able to separate from the spirit of the times. This happens always and everywhere and must not be mixed up with identification.

Nowadays we find ourselves, with regard to petty officers, in a development stimulated by alterations mainly in the field of increasing technics [the study or science of an art] which have caused a change of tasks. These tasks are of a different nature than the tasks in their former practice in their former capacity as superior of groups and thereby as mediator and generator of discipline. It is not possible to ignore sometimes that under these new aspects there are petty officers who try to withdraw into the position of a technical expert or skilled specialist only. This problem cannot be excluded from officers, especially because theoretical education carries more and more weight. The integration of leadership as one component of education must be observed emphatically. An officer must have a sense of his people and their complexity in modern times. After all, a military leader has to be a competent,

personality-oriented person in the practice of this profession, and not a theorizing matter-of-fact person.

3. Although the significance of "leadership" has been recognized for years, its integration was repeatedly contested. To me it seems there is no doubt that "leadership," "training," and "education" make *one* unit, in which one of these components must not be separated from others. It is this unit only which guarantees the object aimed at: readiness for action of a naval unit on the combined basis of technical knowledge, efficiency, and spirit of personnel.

4. As these principles existed in the Kriegsmarine and worked rather well during the war, it has been suspected again and again that this state of affairs has been achieved by ideological indoctrination. Therefore it may perhaps be useful when I state in this connection that I did not experience one hour of [state-sponsored] "ideological" education during my career in the navy. Therefore it has to be stated that morale and fighting strength at that time already resulted from the quality of leadership. The standard spirit and readiness for teamwork was enormous.

Apart from the question of personal contact with the crew of a naval unit, it seems to me that under the circumstances of today, willingness and obedience must be obtained more than in former times by means of conviction in order to get a sincere or upright readiness for operation as the expression of an inner attitude.

5. There have never been written so many books and articles on leadership, morale, and team spirit as there are today. In former times the daily practice was always of more importance than writing and thinking on it. We are living today in a time of almost inundation by theoretical contemplations, considerations, and analysis of almost every activity which, in former times, have been considered an uncomplicated part of naval life.

I do not declare myself against it, because the development we are experiencing calls for tribute. But I may quote a remark from a German essay, which should dispose a naval officer for thoughtful reflections: "The excessive application of science impairs the practice-oriented thinking and acting."

MCPON Sanders: *If the leader accepts responsibility, then the majority of the time he will hold his subordinates responsible, which in turn will help to develop leadership traits in subordinates. Keep in mind that when a leader does accept responsibility for something, he is also making a commitment for his subordinates, which may sometimes be good or bad from the point of view of the subordinates. If you're just coming back from a cruise and you jump out and make a commitment for your subordinates when they were expecting to go home to their loved ones, you may not be properly exercising responsibility if you have not considered the plans of subordinates. This problem can be avoided by doing as much maintenance and repair as possible at sea, so that port time will not require the full crew on board.*

Chapter 7. Study Guide Questions

1. According to Admiral de Cazanove, how does a leader show that he or she is responsible?

2. What does General Rice say about being given an unpleasant assignment?

3. How can a leader develop a sense of responsibility among subordinates?

4. What is the relationship between delegating and assigning responsibility for a task or mission?

5. When will failure of a junior be regarded as unavoidable by a senior?

Vocabulary

propensity
hallmark
psychic income

Chapter 8.   Authority and the Naval Officer

Adm. Charles Larson points out that

we need to develop an officer corps that, within the triad of responsibility, accountability, and authority, has an ethical code that ensures that officers always do what is right and always use authority properly. Junior officers need to see that their leaders always act in an appropriate, proper, and legal way, so they will have confidence, when orders are issued, that those orders will be legally constituted and that authority will be properly used. We need to have an organization whose members must react to orders. In my view, the only reason for not obeying an order would be if that order was illegal or immoral. If we build a foundation that ensures that we do have a code and a corps of people who are honest, who react properly, and who never abuse that authority but use it for the good of the unit and for the mission, then we'll never have an organization where people have to force that authority on people or say "You have to do this because I'm the boss." It will be an instinctive reaction to follow their leadership, and it will be an environment where people know that the right thing will be done.

The ability to exercise authority properly is one of the most important qualities of leadership. Authority can be broadly described as the freedom granted by seniors to allow individuals to command or influence the behavior, thoughts, or opinions of others. Proper uses of authority in the military are based on officers' ability to "show in themselves a good example," support the establishment of the Naval Service and all that it stands for, support and enforce the laws and regulations, and "promote and safeguard the morale, the physical well-being, and the general welfare of the officers and the enlisted persons under their command."

The authority of an officer over subordinates rests not only on the officer's legitimacy due to his or her position, but also on legitimacy due to acknowledged high moral qualities, strength of character, and capacity for leadership. General Barrow believes that authority must never be abused.

Having authority, you are able to compel people to do whatever it is you ask them to do, and it's very easy to abuse that by asking them to do things they shouldn't do, or that don't need to be done. It is demeaning to them and may even be illegal. Certainly you shouldn't ask someone to do something just because you have the authority to more or less compel him to do it.

For an American citizen, authority can only be legitimately exercised in a spirit of liberty. An admiral in the Naval Service expressed it this way: "The American philosophy places the individual above the state. It distrusts personal power and coercion. It denies the existence of indispensable men. It asserts the supremacy of principle." Admiral Uchida suggests that

an officer's authority should be of such magnitude that subordinates are devoted to performing their duties as he orders. An officer's authority will be extended when he commands his subordinates from the most important spot in the battle line, following the noblesse oblige doctrine himself.

An officer sometimes must, whether it is wartime or peacetime, give subordinates difficult directives to accomplish. The officer should do everything he can to assist them until they understand that the directives they received are in the best interest of their mission and until they feel honored to have been selected as the task members.

However, it can also be argued that authority, properly exercised, reinforces discipline and enhances the necessary military virtue of obedience. The great naval theorist and writer Alfred Thayer Mahan makes this point: "The duty of obedience is not merely military, but moral. It is not an arbitrary rule but one essential and fundamental; the expression of a principle without which military organization would go to pieces and military success be impossible."

Obedience in a positive sense, then, means the uniform adherence of everyone concerned to a set of rules, customs, and conventions that balance individual freedoms with protection of the interests and requirements of the unit or society as a whole. Thus, discipline and obedience result from the proper exercise of authority, which enforces adherence to the rules. Both discipline and authority are essential for the orderly and effective coordination of effort toward shared goals. The common goal of every naval officer is that of getting the "people to do the Navy's job effectively."

THE BASIS OF AUTHORITY

The authority vested by law in the military leader is very real. While authority is inherent in a leader's position relative to subordinates, in practice it is based on other factors as well. As noted earlier, it is derived from the moral qualities and character of the leader. It is also based on the leader's personal abilities, professional knowledge, and humility—recognition that the leader does not and cannot know everything that subordinates know.

The authority of a military leader is strengthened by quiet resolution, the willingness to take necessary risks, and the moral courage to accept full responsibility for decisions. A leader must be as ready to accept full blame when things go wrong as he or she is to share rewards with subordinates when things go right. For subordinates, recognizing authority means submitting themselves to an authority that has demonstrated the right blend of fairness, impartiality, and humanity. It is the individual leader who induces the compliance with his or her orders.

ETHICAL ASPECTS OF AUTHORITY

Good leaders have an ethical claim to authority.

The leader who worries only about the personal gains that can be derived from his or her position has no such claim. Rather, the ethical claim goes to the officer who only asks where duty lies and then seeks steadfastly to carry it out. Troops will not follow an officer they know in their hearts to be promoting self-interests at their expense. While they may give their obedience, they will not give their complete loyalty. Without such loyalty, both up and down, a military unit's effectiveness is degraded, particularly in combat or other situations where lives may be at risk.

In the exercise of lawful authority, then, an officer should be a professional but not a careerist. That is, he or she should seek always to do those things necessary for the good of the service and the nation as opposed to what is perceived to be good for one's career.

Admiral Clausen offers the following:

Authority depends upon several values: proficiency, decency of character, convincing personality, mutual esteem, etc. There is always a correlation between authority on the one hand and trust on the other.

If it will be possible to inculcate attitude and bearing (which must be appropriate with our time) into a type of person, and he disposes of the necessary qualities of leadership, I would call such a personality in our language a *Herr*. I am inclined to think that this would be one educational aim. According to my experiences, subordinates are more willing to accept authority from such a type than from any other type one may think of. It would be absurd, if not dangerous, to have a wrong picture in mind. I understand the German word *Herr* to be a type of gentleman who is modest, well-educated, self-confident but reserved, hard-working, and devoted to life and always ready to bear privation for the community, etc. A *Herr* is the result of a profound education, which would be ineffective if this officer at the same time would not work at self-education. These attributes sound, probably, high-flown. They are used in order to outline the contour of this type of person in general.

If this is the type of officer a navy is striving for, it should determine first of all which are the characteristics of a modern *Herr*, because he will always be a reflected image of society. After that begins a detailed education, which must convince the aspirant of the trust put in officers; otherwise, the education remains superficial. The rule applies here again: Theory of education must be directed at real conduct of life; it should be known to each officer where the navy is aiming.

DELEGATION OF AUTHORITY

One time-tested method for a military leader to build strength within an organization is to identify within it individuals of initiative and energy, and then associate his or her authority with theirs. This act makes it clear to others that these subordinates are speaking and acting in the leader's name. To be able to do this well is one of the keys to effective leadership. The ability to delegate authority, to put it into the most capable hands, and thereafter to support these individuals in their efforts is a valuable leadership asset.

Admiral McKee describes how she determines how much authority to exercise over others.

A lot depends on your knowledge of the people who work for you. I think it's very important that you are able to know who is capable of doing what. I keep going back to accomplishing the mission. I guess that's the first thing that we are always concerned with, regardless of what specialized area we may happen to be in, but training is a very important part of that, and that's a responsibility of the leader. You must see that all the people are properly trained to accomplish the job they do, and you must provide them with the circumstances to accomplish that training, and then you dispense the job assignments depending upon who has what capability.

Explicit orders to some people are more called for than they are to others, because if you have an individual who's not well informed or does not have the experience that other people have, that individual needs a little more guidance or more close supervision than others. There are other people you can give an assignment to and you may not see them for a week, and they report back to you and it's accomplished.

Officers who are unable to delegate authority are not leaders. More than evidencing a lack of confidence in subordinates, they are betraying a lack of confidence in themselves and their own abilities. The officer who cannot delegate authority and is mistrustful of all power except his or her own will not be able to command, even in peacetime. In combat or in other difficult situations, that officer's unit at some point will surely break down. The effective leader will therefore be conservative in his or her use of power and will bend every effort to bring about its eventual distribution among the members of the group.

Command is not a prerogative; it is, rather, a responsibility to be shared with all subordinates capable of carrying out the implicit details of orders that may be given in only the most general terms. The leader who is unable to delegate the authority to carry out such orders erodes the flexibility and teamwork that his or her unit needs to win on the battlefield. Admiral Burke, when asked by Professor Montor how much authority an officer should give someone when asking that person to do something, replied:

Give him complete authority, absolutely. If you don't give him authority, don't let there be any bones about what you're not giving him; make it clear, not only to him but to everyone else.

Dr. Montor then asked: "In other words, give him the same power you would give to yourself if you yourself were doing it?" and Admiral Burke replied:

More to the point, if you don't have trust in him, then restrict him, but restrict him so that everybody knows what the restrictions are so that he is not responsible for your restrictions, so that you know who's responsible for what.

THE PROCESS OF AUTHORITY

The personal responsibility of military leaders is as important as the authority that is vested in them by law. The manner and completeness with which they accept and discharge that responsibility indicates to a large degree their capacity for leadership.

It is not easy to get people to subordinate themselves willingly to the command of authority. To assist leaders in accomplishing this, the military has a body of long-standing and refined rules and regulations. These have evolved to produce in personnel military virtues such as accountability, integrity, selflessness, loyalty, and bravery—virtues that can spell the difference between victory and defeat in battle. But rules and regulations are worthwhile only if the authority underlying them is properly exercised and enforced.

To exercise authority properly, leaders must be both demanding and dynamic. Leaders are charged with the responsibility and authority to uphold the rules and regulations by demanding that they be followed, and they must enforce conformance when doing so becomes necessary. The leader's consistent enforcement of authority will serve to ensure that the standards and habits formed in peacetime are predictive of performance in wartime.

Officers can exercise authority in a positive way or a negative way. Authority that is exercised in a positive sense, in fulfillment of the responsibilities inherent in the leader's position, is generally well accepted by subordinates and is supportive of discipline and obedience. Authority that is exercised in a negative sense—that is, applied arbitrarily in order to enhance the leader's personal goals at the expense of subordinates—is often accepted by subordinates, but in a grudging way. In the long run, negatively exercised authority is destructive of discipline.

Authority may be used arbitrarily when the leader has little time—as in combat or emergency situations—to weigh the strengths and weaknesses of a range of possible choices. However, in most leadership situations, authority is exercised in a more predictable and more proven manner.

The best leaders tend to be those who are expert in the proper exercise and delegation of authority. They are exacting perfectionists in regard to performance of duty, maintenance of discipline, and attainment of the loyalty of subordinates. They are authoritarian, but at the same time they constantly evince concern for the personal welfare, interests, feelings, and rights of their personnel. The military leader who strikes the proper balance between exercising authority and taking care of subordinates can expect to meet with success.

The foregoing clearly provides the overall view of the Naval Service. The following remarks by Admiral Burke are provided to remind officers that they are expected to think, for it is impossible for seniors to spell out in advance all the actions expected of juniors.

Everyone has authority, but not everyone is willing to use it. Nearly all people in the Navy stay too closely inside their orders. They circumscribe themselves on what they can and cannot do. What is needed quite frequently is for them to go a little beyond. You get into trouble that way if you're wrong, but you can also do some good things if you're right.

Admiral Burke explains why it is important for individuals to take authority and exercise it.

The President is ultimately in command, but he can't know everything, he can't be told everything, he doesn't have time. So what the President has to be dependent upon are his people doing things that they ought to do without even informing him, because he doesn't have time to be informed of minor things, yet he should be kept informed of those things he should know about.

Particularly in a combat situation, not necessarily in battle, if you have to ask for permission, it may be too late. What can the President or other people back here know of your particular situation? Sometimes you have to take authority. But you can't start a war, you can't be the wild life major and start World War III. There are many times when, if the subordinate can make the decision and do a job, he can do it, but if he

has to ask for authority to do it, nobody can do it, because there's nobody who has the authority without a great big confab that takes four days, and then it is too late. Taking authority in certain situations is like handling a ship with the division commander aboard the flagship. The division commander has nothing to do with controlling your ship, the captain is in command of the ship. The captain gives the orders to the ship, but the division commander gives the orders to the division, and the captain of the ship accedes to those orders. But if he is in a torpedo attack or an exercise of a torpedo attack, for example, and the division commander has given a certain course and the captain of the ship sees something wrong, he's got to give orders to the ship right quick. He can go right to an all stop, back full, not always just to avoid collision, but to do the right thing. He's taking authority that is not his, that is his division commander's. Over and over again this happens, particularly in combat itself, where people do things that they aren't authorized to do. Sometimes it's wrong, sometimes it's absolutely wrong and a lot of people die because of it, but more people will die if people never do what they are sure is right and should be done.

MCPON Sanders: *There may be a tendency among weak leaders to use authority improperly. They may try to get things done that will benefit them, and they do so using the authority an officer has. In seeing this happen in the past, I can tell you that no one has been fooled, especially not the enlisted community. We teach the enlisted to follow the rules and regulations and, especially, the orders of our officers, which places a special requirement on the officers not to ask for anything which is either improper or to the officer's personal gain. You have to be very careful to make sure that, any time you use your authority, it's done properly.*

Chapter 8. Study Guide Questions

1. What is one of the most important qualities of leadership?

2. Upon which things does a leader's legitimate authority over subordinates rest?

3. How can a military leader's authority be strengthened?

4. What is one of the keys to effective leadership as regards authority?

5. What is the difference between positive and negative exercise of authority?

6. Who are the best leaders?

Vocabulary

legitimate   demeaning

3

Achieving Effective Communications

Effective officers are effective communicators, expressing ideas well both orally and in writing. Further, they project an image of positiveness; of professional knowledge commensurate with age, rank, and assignment; of self-assurance; and of understanding. These attributes are continuously communicated by actions and demeanor.

Since integrity is one of the most important attributes of leadership, all of an officer's communications, direct and implied, must be sincere. Use of "lip service" and a facade of seeming to do the "right thing" is soon discovered by seniors and subordinates alike.

Effective officers should concentrate on the positive aspects of good communications, but they should be aware that there is also a negative side of communications. The individual who always has an alibi is communicating: sending the message that his or her paramount concerns are self-interest and dodging responsibility. The officer who chronically uses excuses—"Not on my watch," "Mr. Gish's division was supposed to do that," "According to regulations, . . ." "The widgits weren't delivered to me on time"—is not a leader. That officer lacks a sense of responsibility, displays no initiative, and is lacking courage.

Leaders must also communicate an aura of approachability. Subordinates will be reluctant to tell an unapproachable "leader" about potentially dangerous situations because of their fear of harsh rebuke or excessive punishment. This reluctance can lead the fearful subordinate to attempt to remedy the situation on his or her own. The CO must always be kept apprised of what goes on in the command; otherwise, events can occur that are detrimental to the unknowing senior, who has the ultimate responsibility. Likewise, an officer must never hesitate to inform the CO of when, where, and how the officer goofed.

Col. James W. Hammond, Jr., USMC (Ret.), pro-

Source: Adapted from chap. 7 of Karel Montor et al., eds. *Naval Leadership: Voices of Experience*. Annapolis: Naval Institute Press, 1987.

vides this example of an officer who projected an image of both being willing to listen and tolerating no nonsense.

As a Plebe, I put in a dining-out chit for Sunday dinner with a family in Annapolis. Regulations allowed Sunday dining out with parents, roommates' parents, and commissioned officers. (The current sponsor program did not then exist.) The head of the family was a St. John's graduate (in the days when it was a military school) and had a reserve commission in the Army. On my chit I identified him as a lieutenant colonel. The midshipman company commander forwarded the chit and recommended approval. The company officer, Lieutenant Commander G., was suspicious. I was invited to his office for questioning:

LCdr. G.: "Tell me about this officer."

Midn. H.: "He's an Army officer, sir."

LCdr. G.: "Does he wear a uniform?"

Midn. H.: "Well . . . ahem . . . ah . . ."

LCdr. G.: "Mister, you're trying to *big deal* me!" [or words to that effect].

Midn. H.: "Yes, sir!"

LCdr. G.: "Request approved. Any other answer and I would have denied it. Dismissed!"

Colonel Hammond had put himself in the soup, and the best he could hope for was a well-deserved chewing out for a frivolous request. Instead, he received a lesson in leadership that he no doubt had occasion to apply several times over the years.

Colonel Hammond relates an incident in which a division logistics officer did not communicate his requirements to those who could have helped him remedy a bad situation. Instead, he tried to please the chief by doing everything himself.

In the late 1950s, the 3d Marine Division was making an orderly redeployment from the Japanese main islands to Okinawa. The Marines were to construct new camps and facilities while initially living under canvas. Suddenly, for political reasons, the movement was expedited. It became an aggravated rush with the consequent confusion.

One of the worst problems was accounting for the vast amount of supplies. There is a difference between having supplies on hand and knowing what is where. The logistical records showed that the division had ample supplies—but they couldn't be identified for issue.

The G-4 (division logistics officer) was thoroughly perplexed as he tried to sort out the mess. He wasn't helped by the constant badgering of a fellow colonel, the chief of staff. The latter's impatience was accelerated by the fact that he was in the zone for selection to brigadier general. The more the chief demanded of the logistics section, the less the G-4 delegated authority to his subordinates. Just then a typhoon hit, and even the canvas was blown away. A month later, a new commanding general arrived and immediately ordered all the principal staff officers (chief, G-1, 2, 3, 4, division engineer, etc.) to Kyoto on a fool's errand designed as relaxation. They were to meet him later in Yokosuka for the Seventh Fleet Conference.

The first night in Yokosuka, the commanding general assembled his staff at dinner. After dinner he spoke. "Gentlemen, you have been gone a week, and I am pleased to report that the 3d Marine Division is alive and well, despite your absence. No man is indispensable. You must communicate your desires to those who can take care of them and not try to do everything yourselves." (Incidentally, the new CG brought his own chief of staff, who had a better feel for communicating the general's wishes, and the former chief was not selected for brigadier general.)

Colonel Hammond relates that when he was plans officer in Fleet Marine Force Pacific, he had 15 action officers under him, and he suggested that they avail themselves of a "Ropeyarn Sunday," or afternoon off each week to relax.

The work load wasn't heavy and most of them had already spent two tours in Vietnam. (And, after all, everyone already thought we spent all our time on the beach.) Evidently, they didn't take me seriously or were afraid to leave their desks. I had to find a way of communicating, but I couldn't use the general's old ploy of sending them to Japan. I needed another way to communicate. At the next cocktail party, I asked each lady why her husband wasn't taking advantage of a weekly afternoon off in Hawaii. The message

got through; afternoons off were arranged among the 15 officers and, not surprisingly, the quality of work improved.

In an organization as complex as the Naval Service, it is impossible for anyone to know everything, particularly concerning areas outside one's immediate assignment. Therefore, the officer should not try to be the expert in someone else's job, and he or she does not have to answer a question just because one is asked. One of the first things a plebe learns is to say "I'll find out!"—not "I don't know." Even the latter answer, however, is preferable to giving one that is inaccurate.

In addition to being able to reply "I'll find out!" the officer must be able to accept that same answer from a subordinate. If, on the other hand, a subordinate supplies quick but inaccurate answers just to appear to know everything, the officer should challenge the answers the first time an incorrect response is noticed. It will be cheap tuition for the subordinate to learn the lesson early, before the automatic response becomes a habit and he or she is caught in a more serious situation.

One hallmark of an effective leader, in both command and staff assignments, is the ability to write clearly and concisely. There is no place for ambiguity in military service. Staff papers should be models of brevity and clarity. There should not be one unnecessary word or any stilted language. The organization and syntax should convey the message. The reader should have no doubts about what is expected or described. A staff paper should spell out for an action addressee WHAT, WHEN, and WHERE. The addressee should be allowed to determine HOW.

Effective communication is also important because a large portion of a leader's time is spent communicating with others. A leader's day is filled with face-to-face communications with seniors, peers, and subordinates. When not speaking directly, the leader may be communicating through other means, such as memos, letters, and reports (which will be discussed later).

Colonel Hammond describes a singularly effective communication.

During the heavy fighting in the late 1960s around Con Thien, south of the DMZ in Vietnam, a Marine infantry battalion was fighting head-to-head with a NVA regiment. The latter was trying to flank the Marines and isolate them by seizing a bridge to the rear of their position. The battalion commander picked up his radio and in the clear gave instructions to one of his company commanders, "Art, there's a NVA column heading toward the bridge. If it gets there before you do, I'll have your ass!" The Marines held the bridge.

The officer who writes professionally and submits material to one of the several professional military and naval journals is enhancing both career and communications skills. Every time an officer sits down to write, he or she becomes better at it. Writing is especially rewarding if the work is published. The officer is identified as one who is willing to offer ideas to the scrutiny of seniors and juniors alike and to accept the criticism of contemporaries. Writing professionally builds an officer's "service reputation."

Colonel Hammond provides this example, which illustrates where effective communication leads.

Years ago, a Marine major general had a young captain as his aide. The aide was tasked with writing the general's memoranda, letters, and speeches. Almost all of the captain's writing was rewritten by the general and was hardly recognizable to the captain in its final form. One day after the general had given an outstanding speech, the aide quite guilelessly said, "Gosh, General, I wish I had written it like that."

The general replied with a smile, "If you had written it like that, then you would be the general and I would be the aide!" The bottom line is that the major general became the Commandant of the Marine Corps.

Chapter 1.   The Basics of Effective Communications

People cannot be led, save on the end of a rope, if they cannot understand what the leader says, means, or expects. Clear direction is a critical dimension of organizational climate. When people understand the mission, values, standards, and expectations of the organization, they can do what needs to be done. Lack of such understanding leads to false starts, ineptness, and discontent. Admiral Holloway suggests that

in communicating with seniors, the officer should try to let them know what his plans are, but only up to the point where his plans are relatively firm. The officer should not continue into an area where he is not on firm ground, because conjecture is not helpful to the leadership. If he does discuss the distant future, where conjecture comes into play, the officer should offer rational appraisals of what his probable future course of action will be.

An officer should tell his subordinates what the missions of the organization and his particular group are and, in general, how the officer plans to accomplish them. Subordinates usually are very bright, concerned, and interested, and they should be kept informed.

Generally speaking, however, in communications, the further down the chain of command, the closer to the front line, and the more the troops are involved, the less security a message has. First, the communication systems that are used in front-line units tend to be more vulnerable to code breaking and listening-in by the enemy. Second, junior people tend to talk more than senior people because they do not understand as well the need for security. So, in deciding what to pass down the line, the officer should remember to tell people what they need to know to do the job and what is important for their morale purposes, but he should avoid providing information that, if it got outside the command, could impair his ability or be a detriment to his plans or do harm to his country.

The officer should not communicate his plans for the future laterally to anyone except those individuals who must know. How guarded he must be depends on the officer's assignment. For example, the leader on a mine sweeper must be more guarded in communicating to those in charge of other mine sweepers than a task force commander at headquarters who is communicating to other headquartered task force commanders.

There are many kinds of communications, but the same rules do not necessarily apply in electronic communications and in person-to-person communications. The officer who is talking to his troops at quarters for muster and sort of pumping them up and telling them what a great bunch they are can be more wordy and innovative than a strike leader who is calling back to say that five SAMs have just lifted off. In the first case the officer can afford to use a better rhetoric; in the second case he just wants to get that message out quickly. The same idea applies when a four-star admiral writes a message to the Chief of Naval Operations or the Joint Chiefs of Staff. His communication will be more philosophical than the message of a destroyer skipper to his task group commander telling him what he is and is not going to be able to do in the projected operations over the next 48 hours because of a problem with his number-one fuel pump. The number of people in the chain of command for communication should be reduced to its absolute minimum, but not one less than the absolute minimum. Arbitrarily saying that there can be no more than five people in the chain of command, for example, is not sound policy, because the number of people in the chain of command has to be based on the circumstances. For example, if only five people were allowed in the chain of command, the individual who has the helicopter air-sea rescue unit and may be called into play at any minute to extricate people might be left out.

There should not be anyone unnecessarily included in the chain of command, however. Because someone is a nice person and is interested does not mean he needs to know. He should not be in the chain just for that reason.

Many basics are involved in effective communication, among them audibility, articulation, spelling, and grammar. A long-standing naval tradition is for officers to exceed minimum requirements. Refinement is expected. The way an officer speaks, writes, and thinks should convey an image of an educated person. Much of this striving for excellence stems from a concern for image and credibility, and it is an entirely practical and fine tradition, alive and well among outstanding officers.

EXPLAINING DECISIONS AND SHARING INFORMATION

When power is effectively used, people do not feel like pawns. Rather, they understand and subscribe to the goals and values of the organization and feel empowered, not dominated. When the reasons for orders are not apparent, the likelihood of the orders being disregarded increases. A steady diet of such orders breeds resentment.

Emergencies, when there is no time for explanations and officers must rely on trust, do arise. Trust is usually built through people's discovery over time that officers have good reasons for what they require. For example, there are compelling reasons for the insistence on obedience, respect for the chain of command, cleanliness, and order. These are not arbitrary personal preferences, but if these requirements are viewed as such, personnel will be little concerned about slipups as long as they avoid being caught. When the standards are seen as being vital to the health and survival of the organization, people will work at upholding them without constant prodding. Outstanding officers seem to understand the principle of providing reasons for orders quite well, as the following example of communications illustrates.

Okay, the ship is going to sail over the horizon with twelve airplanes on it. The only defense that ship has in its battle group, in its cruisers, in its destroyers, are the F-14s. So if we're ready, we can protect that battle group. If we're not ready, then the battle group might as well just go back to port and tie up. Now, the Chief of Naval Operations won't hear tell of that. The President, the Joint Chiefs of Staff, nobody would allow that to happen. So it comes down to you. You've got to be on your toes. You've got to be ready.

And they got ready because they understood the mission's importance.

Officers can share information with department heads through regularly scheduled meetings. They can inform their personnel of upcoming events so people can plan ahead. COs and XOs can also make certain that the crew is briefed on what to expect during special evolutions. Even trivial rumors must be squelched, as the transmission of *accurate* information is vital.

Sometimes COs personally brief the crew at quarters in preparation for inspections. Members of the crew are told what is going to happen and how they are expected to behave, generally in words similar to these: "No need for you to be nervous. The house is in order. You are well prepared for this inspection. Try to relax and give honest answers to any questions the inspectors have."

MAKING SURE PEOPLE ABSORB WHAT IS COMMUNICATED TO THEM

When they talk to people, outstanding officers watch the expressions of those in the audience to be sure they are taking in what is being said. Where there is any doubt, officers ensure understanding by making a clearer restatement or requesting that an individual repeat the message in his or her own words. Important oral communications are sometimes followed up in writing, or vice versa, all to ensure uptake and understanding.

The importance of making sure that oral communications are understood is illustrated by the example of an air squadron that was out for muster in a noisy open hangar at the airfield. To make sure that everyone in the squadron could hear him, the commanding officer pulled everyone in close around him, in football-huddle fashion, instead of leaving them spread out and neatly lined up.

TAILORING COMMUNICATIONS TO THE AUDIENCE'S LEVEL OF UNDERSTANDING

Many factors are involved in tailoring communications to people's level of understanding, an advanced skill of leaders and teachers. If a leader wishes to have people understand a message fully, he or she must think about what they need to have in order to do so. A few techniques related to tailoring communications to people's level of understanding for solving communications problems are listed below.

1. *Problem:* Insufficient background information.

Solution: Provision of background information. He [DH] was new to the ship. He launched into the Supply DH about the way he was handling configuration management. He didn't understand the logic of the procedures and wasn't listening to the senior officer. The CO decided to intervene and take him aside and tell him some sea stories about fiascos that led to the SupO's current procedures. Once he understood the background, the DH saw the procedures' merit.

2. *Problem:* Idea too abstract or too theoretical to grasp.

Solution: Provision of a concrete example that connects with the person's experience. In high-stress conditions, people regress [theory]. Take Steven and Tom, for example. They are rocks of Gibraltar. Ordinarily, they wouldn't dream of blaming their mistakes on anyone else. But here was pressure like they'd never seen, and they babbled incessantly about how stupid the other one and everyone else was but themselves. They were behaving like kids [concrete example].

3. *Problem:* Idea too strange or too complex to grasp, or too pedestrian and commonplace to remember.

Solution: Use of a metaphor or an analogy. The lady [the ship] [metaphor] is suffering the usual maladies of middle age. She requires an extra dose of tender, loving care [analogy].

4. *Problem:* Vocabulary and phraseology too exalted.

Solution: Assistance of an editor.

5. *Problem:* Vocabulary strange.

Solution: Avoidance of the use of Navy or Marine Corps jargon with non-Navy or non–Marine Corps audiences.

In summary, outstanding officers make sure that people "hear" them, understand the message, and understand its import. In doing so, they uphold a long-standing naval tradition.

Making Sure Personnel Are Continuously Informed

In superior units, plans of the day (POD) are viewed as a major vehicle of communication. They are always thorough, and they include the long-range view. Through plans of the day, personnel are alerted to upcoming events and to what they are expected to do. Their attention is drawn to important issues. One plan of the day reported that a man had fallen down and broken his leg because a hatch had not been properly secured. The POD did not point out where the accident had happened, so that all hands would be careful.

Another aspect of keeping the troops informed is the indoctrination program for new troops. All units must have these programs, and the effort and time put forth to make them effective is worthwhile. In an effective indoctrination program, commanding and executive officers get directly involved and talk to all new personnel, individually and as a group. Indoctrination programs must be timely; only in average units does the indoctrination program take place three or four months late. In one poorly organized program, the instructor had to ask people in the room for the latest information, and the commanding officer who was supposed to attend had gone flying instead.

Cross-Compartmental Communication

Cross-compartmental communication is essential. Operations officers must talk to maintenance people; the engineering department head must coordinate with the other heads of departments; Marine Corps battalion staff members must coordinate effectively with company commanders. Without telling anyone, one engineering department head arbitrarily cut off a communications line to the commanding officer. After considerable inconvenience to the commanding officer, the operations officer finally tracked down the problem and restored the cross-compartmental communications.

Maintaining Contact with Personnel

As part of their emphasis on communication up and down the chain of command, leaders should regularly walk around their units to learn what is going on. This is a *planned* activity, not something that is squeezed in during spare moments. In this way, leaders gain a sense of their unit as a whole, instead of focusing excessively on details. An officer on a ship who spends too much time below deck, focusing on individual activities such as the installation of a pump (which will only make the crew nervous and impede progress), is missing the opportunity to monitor all activities in an unobtrusive way.

Officers should emphasize the importance of communication up the chain of command. During tours of the unit or spaces, officers are alert to working conditions, individual performance, and opportunities for individuals to say how they think things are going. This approach is not an invitation for circumvention of the chain of command or confrontation; rather, it allows individuals to communicate directly, and it also provides a chance for officers to express their interest in the welfare of their troops.

REMAINING AVAILABLE AND VISIBLE

Outstanding officers "manage by walking about." Yet they do this only partly to keep informed and monitor what is happening. Another very important part of walking about is to show interest, concern, or appreciation. Many senior officers make an effort to get to know people's names and something about them. On their rounds, these officers can make statements such as: "Seaman Jones, how's your mom doing? Has she recovered from her operation?" or "Private Smith, I hear you're getting ready to ace the essential subjects test." Overall, the walking about is upbeat; it is not a fault-finding mission.

Officers can show interest by sitting in on training, observing drills, and stopping on a stroll to watch an evolution: "Carry on; I'm just watching. Never saw it done quite like that before." During preinspection drills, one ship's CO stood up on the bridge spotting people putting out extra effort or doing an especially good job. The CO would get on the speaker and single them out: "Nice job, Jane Smith. . . . Can't see Tommy Jones, he's moving so fast. . . ." Several officers noted that they genuinely enjoyed these strolls and that the strolls paid off considerably.

MCPON Sanders: *When communicating with the crew, I would first recommend that the officer discuss what is going to be said with the senior enlisted,* *and let that individual provide some of his own thinking and arrange the meeting with the troops. In this way you will know that what is being communicated is consistent with the ideas of the senior enlisted, or, if there is disagreement, that can be worked out ahead of time so that you provide a unified front to your unit. Going through the chief serves several purposes. First of all, it gets his confidence; he understands that you are working through him, and it builds him up a little bit; second, if you just walk into an area and call all hands up to let them know what you want to talk about, you may be disrupting previously scheduled work. Thus, it's a lot easier to go through that senior enlisted and then speak directly to the crew.*

Chapter 1. Study Guide Questions

1. What is critical to the accomplishment of a unit's mission?
2. What does Admiral Holloway say about the security of a message?
3. What are the basics involved in effective communication?
4. What occurs when power is effectively used?
5. What are five techniques for tailoring communication to people's level of understanding?
6. What does MCPON Sanders recommend an officer do prior to communicating with the crew?

Vocabulary

conjecture
subscribe
empower

Chapter 2. Producing Effective Oral and Written Communications

People communicate to let other people know their thoughts. In the military, effective communication is particularly important, because mission accomplishment usually depends on everyone acting with the same knowledge. Effective communication is based on five basic principles. These are knowing what to say, whom to say it to, when to say it, and how to say it, and having the responsibility to improve the communication network. Therefore, it is advisable for an officer to determine in advance whether people are likely to understand what he or she has to say as it was meant to be understood. This saying makes the point well: "I know you believe you understand what you think I said, but I am not sure you realize that what you heard is not what I meant."

People learn through experience to evaluate how well they have communicated what they intended to communicate. By a lack of response or a failure to act, the people receiving the message indicate that the sender did something wrong. The effects of the feedback will cause the sender either to continue with a certain style of communication, when the feedback is positive, or to change it, on receipt of negative feedback. The feedback is intended to help the process and must be specific and well timed in order not to overwhelm the receiver.

An important lesson in communicating, then, is that the sender of the message has the responsibility of ensuring that the receiver will *understand* what is said. In the World War I Battle of Dogger Bank, faulty messages were given, or messages that were sent were never received. A flotilla of ships failed to carry out its mission because of ineffective communication. It is, therefore, prudent for an officer to check with someone in advance of communicating, if there is time, to determine whether the message is likely to be understood by others. Both officers and enlisted with whom the officer is associated are appropriate persons to review the message, for in many cases individuals in both groups will have to act on the information. On the other hand, if comprehension of the message requires a specialized background, it is appropriate to have the communication checked by someone with the appropriate specialized knowledge. Consulting the various experts in each separate field can give a person a greater understanding of how to communicate better. Also, availability and cost can be taken into account.

Before having the material reviewed by others, a writer should prepare a draft, keeping in mind the receivers' knowledge and interests. He or she should revise ruthlessly and ask whether the contents are clear and accurate. A good technique is to try to find fault with the draft by quarreling with the need for every paragraph, every sentence, and every word. In order to avoid using terms others may not understand, the writer could test his or her language by assuming the role of the receiver. "Just Plain English," written in the Office of the Chief of Naval Operations (OP-09BR, Washington, D.C. 20350), is a valuable aid. As an example of the changes in communication style over the years, consider the old and new preambles to Executive Orders.

Old: By virtue of the authority vested in me by the Constitution of the United States of America, and as President of the United States of America, it is hereby ordered as follows . . .

New: As President of the United States, I direct . . .

Once a draft is written, the officer should read it aloud. It is a good "rule" never to say anything by mail that would not be said in person. For most people, good writing means good rewriting. Revision is worth the effort. A single naval letter is likely to be read by many people as it goes up for signature in one activity and down for action in another. The officer should work to help the many people who must read his or her writing. If the writer does not sweat, the readers will.

When consulting with others regarding the communication, the officer will find that they will probably be able to help by checking to see whether he or she gets quickly to the point, has enough but not too many references, states rules before exceptions, says who does what, gives examples for difficult ideas, and answers likely questions. Until Murphy's Law is repealed, an officer must write material that cannot be misunderstood.

To be effective leaders in today's Naval Service, officers must become accomplished speakers. They must be able to present facts clearly and concisely to individuals and to groups of varying sizes in both off-the-cuff and formal situations. Practicing the delivery of a speech in front of others will help an officer get feedback about whether he or she has the following traits of a good speaker:

1. Has knowledge of the subject
2. Is adequately prepared
3. Is up-to-date professionally
4. Is poised and self-confident when speaking
5. Has a natural delivery style.

Naval officers are encouraged to seek out methods of improving their delivery of information. Writing for publication and speaking to civilian groups are effective means of improving communication style and ability, and by consulting others during the planning phase of a communications effort, officers will find that they are better understood, that things go better, and that others look forward to receiving their messages. Admiral Holloway makes the point that

juniors should be asked for inputs in areas where they are qualified to give them. The officer can talk to a young pilot, for example, and ask him if there is anything that would speed up his check-in procedures coming back from the beach. The pilot might ask why he has to check in with the Tom Cat, and maybe that could be changed. The point is that the officer is asking the pilot for advice in an area where he is competent to respond. Giving people the opportunity to comment in areas where clearly they are not competent is almost always a waste of time.

In writing for publication, an officer should address subjects on which he has some expertise. Rather than outline his ideas on how the Pacific Fleet should be run, he should write on a subject he knows from first-hand experience. He will in his way be making a valuable contribution.

Chapter 2. Study Guide Questions

1. Upon what five basic principles is effective communication based?

2. What are five traits of a good speaker?

Vocabulary

Murphy's Law
feedback

Chapter 3. Determining the Purpose and the Audience for the Message

An effective officer continually works to develop good communication skills, for lack of clarity in communication is a problem even the most effective military leaders must struggle with. For example, General of the Army Dwight D. Eisenhower stated, "All my life I have been an incorrigible reviser of written material. Whether I dictate a draft, or a draft has been prepared for my signature, I find that I have almost never said exactly what I wanted to say, in the way I hoped to say it."

Effective communication up and down the military chain of command is essential to mission accomplishment. Admiral McDonald points out that "if orders are misunderstood, it may be that someone along the line didn't know how to communicate, and that the senior in the chain hasn't put enough emphasis on transmitting orders." Communication is a key link in execution and integration of effort. Before a leader can issue a crisp and clear oral or written order, he or she must know exactly what the goal of the mission is, and what needs to be communicated to whom in order to accomplish that objective.

Once the purpose of the communication is clear, the leader must consider who will be receiving the communication. Seniors receiving the communication will usually voice any confusion they have, especially if the situation as described in the message is different from the receiver's assessment of the situation. It must be emphasized that communication is not just a downward process. It goes both up and down. We must communicate with our seniors just as often as with our juniors, and they must communicate with us.

Subordinates are less likely to indicate confusion or disagreement, so the leader must take steps to ensure that they understand the message. Secretary Webb states that

a leader must be able to communicate to his or her people with a sense of presence and self-confidence. The form that communication takes depends on several factors, including the setting and the leader's personality. Communication can be written, verbal, or physical—such as posture or standing up in combat. But the leader who understands, yet cannot communicate to his troops on a level they will appreciate, is probably doomed to fail.

"Mobilization of minds" is a phrase that, in modern warfare, suggests that wars are fought with words as well as with weapons. An order must be understood by the lowest command level tasked with carrying out the mission or objective. The military leader, then, must know the capabilities of followers before issuing an order. The added stress in the heat of battle or in a crisis demands that words be carefully chosen to avoid confusion. In battle, a missed, ignored, or misunderstood communication can result in poor transmission of ideas, thoughts, orders, etc. With respect to writing, many supervisors or seniors believe they can communicate effectively via messages, command letters, bulletins, memos, and other means exclusively.

In the World War II Battle for Leyte Gulf, one of the main U.S. Navy leadership problems was the lack of timely and clear communications between the Commander Third Fleet (Adm. William "Bull" Halsey) and the Commander Seventh Fleet (Vice Adm. Thomas Kinkaid). The communication problem, compounded by messages obscurely worded, misinterpreted and, in one instance, accidently corrupted, resulted in Halsey and Kinkaid basing several decisions on misinformation. Misinformation is an even more important factor today, as the enemies of the United States actively try to confuse and disrupt military communications efforts in times of crisis. Again, getting "back to basics" helps avoid "slips of the tongue."

In war, self-distraction and miscommunication can quickly defeat an officer. Adm. Chester Nimitz, Commander in Chief Pacific, wrote a letter to his commanders in World War II:

There are certain psychological factors which have fully as much to do with safety at sea as any of the more strictly technical ones. A large proportion of the disasters in tactics and maneuvers comes from concentrating too much on one objective or urgency, at the cost of not being sufficiently alert for others. . . . No officer, whatever his rank and experience, should flatter himself that he is immune to the inexplicable lapses in judgment, calculation, and memory, or to the *slips of the tongue in giving orders*, which throughout seagoing history have so often brought disaster to men of the highest reputation and ability.

MCPON Sanders: *If an officer does not clarify his own thinking on something that he is going to pass down, the orders will become very fuzzy and not understood. The end result will be problems within the command, department, or work force. It is important for an officer to know what he wants to say in advance, formulating his ideas and using the correct words, remembering that what you say should be said in terms appropriate to the audience being addressed.*

1. What must a leader do before he or she can issue a crisp and clear oral or written order?

2. What does MCPON Sanders say about effective preparation for communication?

Vocabulary

incorrigible
assessment

Chapter 4.   Getting the Message Across

A successful communicator understands that people are different and that people respond not only to the carrot-and-stick methodology, but also to ambition, patriotism, self-doubt, and a host of other emotions. General Rice suggests that

> one way an officer can ensure that his intended message has been received is to look at the results. If he has given his orders correctly, if people understand him and know what he wants, they will provide feedback indicating that. When an officer has given guidance, and the paper he gets back indicates that the staff has achieved nowhere what he wanted them to achieve, then obviously something is wrong: either they did not understand or the officer's guidance was poor. In this case, the officer should rethink the way he communicates.

Setting the example is a basic way to communicate. If the officer expects people to look sharp and be sharp, he has to look sharp and be sharp, and he will readily be able to tell whether people have received his message.

An officer must know his job before he can expect someone else to know theirs. That does not mean a unit commander has to be a mechanic or has to know everything there is to know about an engine, but he should know some basic things about the engine. The mechanic gets paid to do the detailed work, but the officer should have general knowledge of what the mechanic is supposed to do, because otherwise he will be unable to supervise his subordinates and check what they are doing. The machine gunner may not know the tactics involved, but he knows that machine gun; the officer has to know how that machine gun should operate and where it should be placed to provide the most effective fire. So an officer needs to have a general knowledge about many things.

After every briefing it is a good habit to ask if there are any questions. There are many ways to do this, but it is important to do it in a way that lets the troops know they are free to ask questions. Many people are afraid to ask questions because of the reaction that sometimes elicits, so the officer must establish a reputation as someone who genuinely wants people to ask questions or to ask for a clarification.

Achieving effective communication in the military presents special problems. Because service personnel face extreme risk, injury, or death and therefore may have to be replaced, sometimes continually, there are important reasons for treating them uniformly and, at times, mechanically. Each person's responsibilities and duties must be delineated clearly and understood both up and down the chain of command. Admiral Hayward notes that

> in communicating, an officer uses the chain of command, transmits as clearly as he possibly can, and meets frequently with his senior officers. He develops policies with their advice so that when the policy is ultimately distributed to the chain of command, he knows that it is starting out understood. An officer can only hope that the chain of command will keep transmitting the policy clearly. The chain of command is the right way to get the word out, from any size unit on up. Good communication is a requirement of good leadership. Some people are naturally very effective at it and some simply are not, but it is one of those skills that a leader must acquire. He must learn to write and to speak with sincerity and objectivity. Eyeball-to-eyeball contact is part of effective verbal communications.

Admiral Hayward explains how his action with regard to drug use was an aid to, not a sidestepping of, the chain of command.

> I've stressed the importance of communicating through the chain of command as a traditional, proper way for the Navy to get things done, and let me suggest that even in this instance the chain of command was employed. I used the television two or three times in four years, in each case for circumstances I thought were out of the ordinary by a long measure, and therefore in order to address the issue we went at it with several tech-

niques, of which the television communication device [video] was one.

In this instance we were talking about the drug issue. After two years of looking upon it very hard I saw no results and I finally got to the point where as the Chief of Naval Operations, with the responsibility to the nation for fixing this problem that couldn't be fixed without taking extraordinary measures, I had to take those measures. Those other individuals in the Navy responsible for helping the CNO develop a policy worked together, and we spent about six months putting into place each piece of the pattern. We met with the master chiefs of the Navy, we developed a program with their support, we got the fleets involved, and finally all the three and four stars. My message to the fleet, while it was transmitted obviously in a personal sense from the CNO to every individual in the Navy, was literally handed down by the Chief of Naval Information to each command, and it was up to each command to use it as they saw fit. They didn't have to show that film if they didn't want to, or they could use it if they thought it was an important tool, and they used it very effectively. So it certainly wasn't a device to subvert the chain of command, it was a device to assist the chain of command grapple with a very difficult problem. I thought it was very encouraging and one of the finest things that the Navy has done for our country.

BARRIERS TO COMMUNICATING EFFECTIVELY

Effective communication is essential if plans and programs are to be successfully translated into action. The fundamental difficulties in communication arise from two assumptions. First, on the part of the speaker or writer, is thinking that others understand what is stated as he or she intends it to be understood. Second, on the part of listeners or readers, is thinking that they understand what has been said as it was intended to be understood. Because we rarely test these assumptions, we are rarely sure that we are truly communicating. René Descartes left us with his rules of logic of speaking:

1. Discard everything except that which you know to be true.
2. Break the subject down into as many parts as possible.
3. Start with the easier parts and progress to the hardest to understand.
4. Summarize.

In other words, you are trying to persuade the person to think like you on an issue, and that is where the true work will begin, when speaking to people; or, to put it in the words of A. H. Leighton, "Man acts in terms of what he perceives, and what he perceives must pass not only through his eyes, ears and other special senses to reach his consciousness, but also through the dark and iridescent waters of his belief."

The higher officers go into management and leadership positions and the more authority they wield, the less they are forced to listen to others. As individuals gain authority, they sometimes develop a lack of patience in listening to subordinates, yet their need to listen is greater than ever. A deaf ear may be the "first symptom of a closed mind."

The farther officers get from the firing line, the more they have to depend on others for correct information. If the officer corps has not formed the habit of listening—carefully and intelligently—officers are not going to get the facts they need. The ability to listen is also an important ingredient in successfully transmitting the message. Lessons in delivering orders can also be found in the historical practices of old warriors like Napoleon and Marshal Ferdinand Foch: It is recorded of Napoleon, the most autocratic of men, that he never gave an order without explaining its purpose and making sure that this purpose was understood. He knew that blind obedience could never ensure the intelligent execution of any order. Marshall Foch, in his Principles of War, makes the same point in his distinction between passive and active obedience. "Command," he says, "never yet meant obscurity." Active obedience implies initiative, and intelligence in the exercise of this initiative requires a knowledge of how the specific objective fits into the general plan.

THE EFFECT OF OVERTONES

In both written and verbal communication, it is important to be mindful of the overtones of the message. In verbal communication, the implication of "body language," the sending of additional signals to the receiver through gestures or posture, is well recognized. Additionally, the tone of voice, manner, and pace in which words are spoken can sometimes convey a message to the listener far different from that intended, and far different from the words spoken. Conversely, officers may intentionally use this technique to color the message; however, it is more professional and honest to communicate thoughts clearly and directly, without ambiguity, than to risk miscommunication.

Paul Soper, in Basic Public Speaking, says, "An effective speech is perhaps one-fourth or one-third what your listeners see. They see you: how you stand, the look on your face, how you walk, what you do with your hands—what you are. They size you up, curiously and candidly, favorably or unfavorably. They will decide, without knowing it, whether they are going to like you and whether they are going to accept what you say. However important physical appearance is, it will not take the place of the right attitude. In fact, as the speaker gets into his talk, he may cause his listeners to forget his appearance."

Communication is a two-way street, and listening is a major part of communication, but individuals frequently receive only the words that others utter or write; they have little understanding of the feelings and values that underlie the statements. As a result, an individual often misses the main points another person is trying to make, as in this exchange between a department head and a division officer.

DEPARTMENT HEAD: Jim, you're filling out your PMS forms all wrong. Didn't you read the notice? You're still doing it the old way, and that's what we are trying to get away from. Now the division is all fouled up. You'd better fill out the new forms and get your PMS up to speed.

Notice that the department head assumes that because the instructions were written and distributed, they were properly interpreted by the division officer. He assumes that the notice, which was, in fact, misunderstood, is clear. Notice also that he expects bringing the notice to Jim's attention will make the situation clear to Jim. He was totally unprepared for what followed, however.

DIVISION OFFICER: Commander, this division has never been in trouble because of PMS. We've been doing just fine. Why did we change forms in the first place? I've been in this division over a year without problems. The previous method has worked just fine.

Notice that the division officer is responding to what he thinks is a personal criticism of him and how he runs his division, so he replies in a very defensive way.

DEPARTMENT HEAD: That's the trouble with you, Jim. You only think of yourself. I've made this change in the entire department. You're the only one complaining. As long as I'm the department head, we'll do it my way.

Now the department head sees the division officer's defense as an attack on his authority and responds blindly in terms of his own feelings.

This is a classic example of a breakdown in communication. Neither understands what the other is trying to say. Each is judging the other's statements in terms of his own attitudes and objectives. Little

attention is being paid to the emotions the other is experiencing and expressing.

An officer who wants to improve his or her ability to communicate must take account of the other person's position and try to see the other point of view.

One key to effective verbal communication is to ascertain the characteristics of the audience in advance. Then the leader must adjust his or her speaking style and the message to fit the audience. He or she cannot necessarily communicate effectively, either up or down the chain of command, if the same words and style of speaking are used in both directions. Admiral Carney cautions that

when an officer delivers an order, he should observe the physical reaction as well as the verbal action it elicits, and he should pay attention to how the verbal response is given as well as to what the verbal response is. In other words, how something is said is as important as what is said. I was very aware of the negative attitude of one of my men, and my leadership helped turn the man into a fine sailor. I think you can sense a lack of sympathy or you can sense hostility. I had a sailor when I put the *Denver* in commission, he was from South Philadelphia, he was a troublemaker, and he was a bad actor. When we were crossing the equator and played a fire hose on his back, he came up with a knife and went berserk. He was overpowered, and I put him in the brig. So after he had served his bread and water, which in those days I could administer without the code of military justice interfering, I brought him back to mast, and before he said anything, I said, "I know exactly what you're thinking, and you're going to go back and do some more."

Well, I had a highly intelligent, loyal, first-class steward, and he went down on his own initiative and he talked to this bird and he explained to him that you can't have thirteen hundred people living elbow to elbow, cheek to cheek, on ship and have anybody who doesn't play along with the community rules. He accepted the steward's admonition, which he wouldn't have accepted had I used the very same terms to him. So the message came back that he'd like to see me. I brought him up to mast, he said he'd been wrong, he said the other man had talked to him and explained to him how he'd been wrong and that I wasn't going to have any more trouble with him. I said, "I'm going to tear that out of your record, but I'm going to keep it close by, and if you're sincere about what you say that'll never show up in your record, but if you revert to your bad habits, I'll make sure it is in your records." He said, "No, Sir. You're not going to have any more trouble."

Time went on, we got out to the South Pacific, and one day we were under an air attack and he had a job in the forty-millimeter gun bucket right under the port side of the bridge, with a steel shield around it. This fellow came in strafing, pretty well peppering us, and the gun crew, quite within their rights and their instructions, ducked behind the steel shield. Except this one bird who was a loader, and he kept both those guns going with the ammunition that was in the magazines. I went on the loudspeaker to the ship and I told them about it. Well, you know from that day on, this troublemaker was the most loyal and valuable sailor I had on the ship. If anybody deviated from ship's pride and policy, he was right down their necks and it was a perfectly remarkable transformation.

An individual may be sending a message, but there is no communication unless the other person receives the message. What is said is important, but how it is said may be more important. Voice and inflection convey more than just words; they also denote enthusiasm, concern, elation, gloom, and anger. So, the officer should decide in advance how to communicate. He or she should use the positive approach when speaking. For example, when seeking to gain approval of a new idea, the officer should not begin by criticizing the methods currently in use.

In written communications, unclear or improper sentences will cloud the message. Additionally, the manner in which the reader is addressed can greatly influence his or her perceptions and subconsciously create prejudice toward the subject matter and the communicator. In the following example, the officer is unmindful of the overtones that are present in the memorandum.

MEMORANDUM

From: B Division officer
To: B Division

Subj: LIBERTY

1. The cleanliness of B Division spaces—or lack of it—has been repeatedly brought to my attention by the XO following weekly zone inspections. Despite my fervent attempts to instill pride and professionalism in this division, the situation remains unsatisfactory. Therefore, by order of the XO, liberty is curtailed until all spaces meet established cleanliness standards.

2. The CO and Chief Engineer also know about this, so appeals to the Command Master Chief will be fruitless.

Although the division officer is obviously trying to motivate the division to clean the spaces, the memo also contains coercive measures. The underlying message is that the officer has lost control of the division and probably has little hope of regaining it. Even with the XO's decree that liberty is curtailed, the officer may not obtain the desired results over a sustained period.

The example also brings to light another fundamental precept of issuing orders; that is, the officer should be sure his or her language is understood. An officer should avoid "talking down" to the troops, but the officer who uses phrases such as "fervent attempts to instill" may not be understood by all members of the division. In addition, poor sentence structure detracts from any intended message.

Chapter 4. Study Guide Questions

1. What does a successful communicator understand?

2. According to General Rice, what is a basic way to communicate?

3. What are two assumptions that can cause difficulty in communication?

4. What are Descartes's rules of logic of speaking?

5. According to Admiral Carney, what should a leader do when delivering an order?

6. What do voice and inflection convey?

7. What is important about the manner in which the reader is addressed in written communication?

Vocabulary

ambiguity
inflection
fervent

Chapter 5. Communication: A Two-Way Exchange of Information

By definition, communication can take place only if there is a two-way exchange of information. This implies the existence of a system in which senders and receivers are able to exchange roles. A good leader must be a good listener, and he or she can be, simply by applying the principle of do unto others as you would have others do unto you. Sending and receiving functions are not used equally by all parts of the system. The amount of time spent in sending or receiving is a function of the purpose of the level in the chain of command that is being served by the communications system. If any component of the communications system is unable or unwilling to transmit, communication will not occur. Conversely, if any component is unable or unwilling to receive, communication with that component becomes impossible.

To help ensure that the listener will want to take on board the message given, Admiral Larson suggests that

the first thing is to understand your audience. You have to understand their knowledge base and their interests and you have to talk right to them. Communicate with them honestly, openly, frankly, as if you were just talking to them one on one. Don't pontificate, don't strategize, don't talk in the lofty clouds, but speak to them and speak in a way that builds their knowledge base and involves them in what you're trying to do. Have a common frame of reference, have common interests. Now this will get difficult sometimes, when you're talking to an audience about things that may be controversial, that may not be easily accepted. On my ship and in my division, when I was a division officer, I used to involve my troops in the planning process. Before policy was formulated I would allow the dissenting views and all of the opinions to be gathered together. We would debate during policy formulation and then, once policy was formalized, I expected everybody to get on board once I had made my decision as a division or as a commanding officer. I expected them to get on board and support me firmly as we carried out our mission. If you are going to have to go before the whole crew or the whole division, it is useful to have that private policy planning session first, give that private explanation to some of the senior petty officers and junior officers, so that, when you go into that public forum with a controversial item, you will have their support immediately because they will understand where you're coming from, and then that support will capture the whole crew. But I think the important thing, I guess the bottom line, is just honest communication, as if you were talking to a friend or a contemporary, and talk to them, not down to them.

"Communications" is usually discussed in the context of communication between individuals: one-on-one, one-on-several, or one-on-many. Different techniques or approaches are effective in each of these three situations. For example, although a one-on-one private meeting and a one-on-many public oration both employ oral communication, complex ideas and problems are best discussed in detail in small groups. The spoken word becomes less useful, in a classic communications sense, as groups grow larger, such as when a general announcing system is used to address an entire organization. In this instance, communication exists only in a special sense, and only in one direction, because the receivers are not permitted to become senders except after the fact.

THE COMMUNICATIONS CLIMATE

The midshipman or young officer who understands the factors affecting the communications environment has an advantage over those who do not understand them. It is important for an officer to recognize and compensate for the fact that the chain of command in many instances impedes the flow of communications. Giving orders, sending directives, and establishing policy are easy—too easy. The inexperienced officer in the command structure can get trapped by assuming that communication is occurring because someone in the upper echelons of the organization hierarchy is transmitting frequently. Put another way, those in command are almost always the last to realize that the organization's communications needs are not being served well because the system lacks a well-exercised bottom-to-top communications capability. The ability to receive feedback is an essential ingredient in any unit or control system. The attentiveness of someone in conversation is an indication of how well a speaker is getting ideas across.

Organizations in which certain elements are always senders and other elements always receivers have allowed their communications system to become subverted. Variations on the "I know it all, I don't have to (and won't) listen" and the "You are too low in the organization to understand the problem, you don't know what you're talking about" syndromes abound in almost all organizations. Where the balance is struck between sending and receiving will govern the effectiveness of the organization's communications. The organization's leaders can enhance or degrade the communications environment by their apparent receptivity, by their response to legitimate stimuli, and by the climate of mutual trust and respect they help—or fail to help—develop.

Left to their own devices, bureaucracies tend to control and stifle rather than enhance or facilitate communications. "No, you can't talk to the division officer" or "The XO is too busy" are heard all too frequently. Turnoffs such as these convey the message that the command is not interested in what is happening on the other end of the totem pole. A positive communications "climate" can be turned off easily, while a negative climate is very difficult to turn around.

SHARING INFORMATION MEANS GIVING POWER

The chain of command concept provides a discipline that regularizes communications, but it also can isolate upper echelons from important information sources. Unregulated or unrecognized bureaucratic filtering deprives leadership of information that is frequently valuable, and the organization must remain sensitive to this possibility.

Knowledge is power, and while knowledge can be acquired from a variety of sources, that which is most valuable (and vital to the leader in an operating environment) is usually found among the subordinates in the organization. What kind of environment causes a person to want to tell the supervisor what is happening? What is it that makes that person want to share information and knowledge (and thereby power) with seniors? Not surprisingly, the Golden Rule operates in the internal communications world, too. What motivates an officer to keep the boss informed will also motivate subordinates to keep the officer informed. This motivation is provided by the knowledge or belief that seniors are interested in learning, that they will appreciate what the subordinate has to offer, and that they will act appropriately on the information passed to them.

General Rice suggests that

to make sure that he is in fact receiving the intended message, a listener should, first of all, listen very carefully. And in order to make sure that people will listen carefully, a speaker must have something to say. Someone who holds meetings for the sake of holding meetings or hearing himself talk will find that people will not listen; they will come and they will take notes, but they will not really listen, because they will be thinking of other things.

An officer must use his judgment to determine whether it is better to ask a question during a briefing, in front of the group, or to wait and ask the question privately, after the briefing. It depends on the individual who is giving the briefing and on the circumstances. When I was in the U.S. Strike Command as a lieutenant colonel, the Commander in Chief's comments were taped. One time I happened to be sitting in there when he talked, and I went back to my office and the major general called me in and gave me direction on what the CINC wanted. I just couldn't believe it, because I had sat there and listened to the CINC, and I knew what he wanted, and it wasn't close to what the major general said the CINC wanted. So, rather than counter the general, I went back and got the tape from the command center and played it back, and I heard what the CINC said, and it reinforced what I thought he said. So I went back to the general, an hour or so went by, and he said, "Well, maybe you're right. Go ahead and do it your way." I did it my way, and it happened to be right. So, you've just got to play it by ear.

I had the time and the opportunity to check my facts before I asked a question, but there are times, particularly in a combat situation, when a face-to-face immediate question is unquestionably necessary. An officer had better understand and know up front what the commander intends, and any commander worth his salt will either reinforce what he said or give the answer that is required. The worst thing an officer can do is just sit and mull. If he does not understand what is going on, he can get himself or someone else killed, simply because he was embarrassed to, or didn't want to, ask a question.

Every organizational level must be able to start the communications process, and while transmitting is usually the first step, receiving and listening to feedback is the second and more important step. The person who cuts off the head of the messenger bringing bad news will also miss out on a lot of "good" bad news. On the other hand, if the organization can see that the command puts feedback to good use, information sources flourish.

UNDERSTANDING AND EVALUATING

Listening is not a trivial task. It requires sensitivity, intelligence, understanding, and common sense. Aside from being sensitive, the listener must also be an astute judge who is able to evaluate the information received. A listener must either know the source of the information or have great faith in the reporter. It is essential to know the reporter's biases and special interests and how well the reporter evaluates the information acquired. Harold J. O'Brien and Harold P. Zelko, professors of speech and communications, offer the following criteria for being an effective listener:

1. Attune yourself to the sender. An attitude of waiting to be sold or convinced may keep you in the dark. You must have a positive attitude of wanting to learn or understand.

2. Try to receive, no matter how poorly a message may be sent. Although a poorly organized talk, a rambling speech, a disjointed conversation, make listening difficult, try to understand.

3. Evaluate and analyze as you receive. People

normally send what they want you to hear or they believe you want to hear.

4. Receive objectively. People tend to select senders who offer compliments or make life pleasant. This discourages the sender who has valuable though unpleasant information. Try to see the message from the sender's point of view.

5. Take appropriate action on what you receive. Receiving implies action.

The leader must know if the reporter is reporting facts acquired firsthand or is reporting "processed" data. If the latter, the leader must determine the attenuation factor. Does each successive layer in the chain put an increasingly benign face on bad news? Sometimes by the time the "problem" gets to a level in which people are capable of providing a solution, it does not seem to be a problem at all. This is a major source of frustration to subordinates who see and report real problems and have to live with these real problems on a daily basis.

Admiral Burke notes that

a subordinate generally will not volunteer very much unless he believes his opinion is wanted. I did a lot of little things to obtain information from my people. For one thing, I had a Navy dinner, a mess dinner, every two weeks. I'd have some junior people, and the people who were connected with surface-to-air missiles, for example, and, if I could get them, one or two people who were opposed to surface-to-air missiles. I started off the dinner with drinks all the way round, and I went back to the old, old customs where the junior man speaks first. So the junior lieutenant would get up and he would make a proclamation of what he thought ought to be done, and he told this before anyone else said anything. No one would ever criticize him or any of the other subordinates directly. They would just talk about the subject. It was remarkable how well this always happened. The seniors leaned over backward to make sure they didn't cut the junior down just because they thought he was wrong.

Another thing I did was, sometimes when I put out an order, I got a flock of junior people, senior people, all the way up and down the line, and I said, "Let me know what happens, what is really done on this thing." So when I put out an order, I checked, maybe a couple of weeks later, as to what was happening here. I called lieutenants in, commanders in: "What about this work that's being done?" When my orders were not followed, it usually was not deliberate. The order was modified a little bit, or they interpreted it a little, or it just wasn't that important. In other words, things

didn't happen the way you thought they were going to happen. I never did anything, I never took corrective action, except through the chain of command, but the information I got every way I could get it. I got information from wives, from secretaries, but I never used it except to check the thing officially.

For example, while I was commander of destroyers in the Atlantic before I became CNO, we were in Newport. Newport's a very easy harbor to get into, it's not bad, though sometimes it gets foggy. But I had a destroyer coming into port with a new captain aboard, and I guess he came from Norfolk or someplace. Anyway, he requested a pilot. Well, I had a rule on the staff and it was that I was the only one who could say no. If anyone asked for permission to do anything, asked for anything, no one else could tell them no, I would tell them no. If the answer was yes, they'd go ahead and do it, but only I said no.

Well, that rule worked as this thing came up. I said, "He shouldn't need a pilot and I'm going out and bring his ship in." "You can't do that," the staff said. I said, "The hell I can't," and so I went and met the ship with a barge and I climbed aboard and said, "I'll take over the ship." The captain said, "Admiral, I can take her in." I said, "No, Captain, you asked for a pilot, you got one. I'll take your ship in." And I took her in. I only had to do that once. I never had anyone ask for a pilot again. That went to the fleet in nothing flat.

A third thing I did was, every night when I was in destroyers, I drove my own car. No one in those days had quarters afloat, you lived on the economy. So on the way home I would pick up officers or blue jackets or whoever on the dock and talk to them on the way home. I kept a notebook in my pocket, and I always asked them certain questions. What ship are you on? How is she? And what's she doing? Everything about her. I caught the name of each man, the ship and the captain's name, and maybe one line. Well, after two or three months I got a pretty good line on my ships, on the gripes. They wouldn't always tell me the truth, though they would sometimes, but I could get an awful lot of truth even when they weren't telling me all the truth. No one would ever say "My captain's a louse" or anything like that, but I'd get a half-damn praise.

I got a lot of dope about captains, so much so that when I was called out to be CNO, I had that book in my pocket, and they said, "You've got to have a flag lieutenant." I said, "I've got one, a junior lieutenant." They said, "You've got to have a commander." I said, "This boy's a good boy." They said, "No, he can't handle it, you've got to

have more experience than this youngster has."
Well, I looked in my book to find out what captain of what ship was best—I had these things by ship—and I went down and found a man, whom I did not know. I'd never seen him, but he had the best reputation from my own book. Everyone liked that ship, liked the skipper and what he had done. So I called his wife at Newport and said, "Is your husband in?" She said, "No, Sir, he's on the golf course." I said, "Would he like to be flag lieutenant to me?" And she said he would. "How do you know?" I asked. "I'll tell him to," she said. I said, "Can he be down here tomorrow morning?" "He'll be there." He was a hell of a good flag lieutenant, still is a very good friend, very fine man. But this came about strictly from asking anyone I found a lot of questions. You could ask other questions: What did you think about this ordeal? What do you think went wrong? Well, they'd tell you something, and mostly it didn't matter much. Once in a while you'd get a pointer and it was damn good, though. What it does do is start people to think a little bit, and they would have something on their mind next time you talked to them, or the word spread that the commander asks his people questions, and so they would be interested in some things. Interest is one thing you've got to get.

The officer must be a perceptive listener, because frequently what is not said is more important than what is said. As they practice and gain experience, young officers will develop a facility for asking "the right question."

Admiral de Cazanove suggests that

it is best if the officer can lead others to draw for themselves the conclusion that is his. The principal way this can be achieved is by ensuring that your people know your thinking and approaches to problem solution. The perfect command is not to give any command, because if your people know exactly what you are going to give as a command, they will take the necessary action without further directives having to be given. Sometimes I used to be asked to ask the admiral what he would say, and I would reply that there was no need to ask him that, and when asked why, I replied that I knew perfectly well what he would say, so why ask him? There is no need to ask the boss whether he wants the job done in the way that will give the best results, for that is obviously what he will want. In wartime this is particularly important, and officers must know both the desires and reasons therefore of their senior, as well as what action should be taken to gain the senior's objectives. You do this because you love your country and your boss. When you have the privilege of feeling this way about your boss, it is extraordinary, for you obey not only because he is your boss but also because you "like" him very much—you respect him, and know that you are being well led. Especially in action this is a great benefit and gives all a great feeling, but even if you do not feel this way about your boss, you will do what he wants because you are disciplined.

MCPON Sanders: *To be a successful officer you must work at being a good listener. As a leader, you should not only listen to your crew but, while giving them your message, see the nonverbal feedback that you are getting and determine whether they are listening and understand what you are saying. When they speak to you, direct your attention to them, for if you don't, they will automatically cut you off. Unless you are ready to listen, you will have fewer people coming to you, and speaking to you only when they have to. Be a good listener, take notes, mental notes at least.*

Chapter 5. Study Guide Questions

1. What principle should a good leader and listener apply?
2. What governs the effectiveness of an organization's communications?
3. What is the best setting in which to discuss complex ideas and problems?
4. What does good listening require?
5. What are five criteria for effective listening?
6. According to MCPON Sanders, what must a person do to be a successful leader?

Vocabulary

component	syndrome
pontificate	attenuation factor
subvert	

Chapter 6.   On the Receiving End of Effective Communications

One method of instruction is for the instructor to tell the class what they are going to learn, present the subject, and then tell them what they are supposed to have learned. Communication can be thought of as a teaching/learning process, wherein one individual has information he or she wants to impart to another individual, who wants to receive the information. Communication does not take place just because the person with a message to transmit has a desire for another person to receive the message, a fact that has not gone unnoticed by advertisers, who try to convey a message to the consuming public in a variety of ways, in the hope that one of the repetitions of the message will get through.

This is not to say that the Naval Service is involved in an advertising campaign, for naval units receiving information want the information and thus will exert effort to make sure they receive it. Officers try to communicate to others in terms they will understand, but they also determine whether the receiver is ready to receive the message and wants to do so.

The ability to convey messages from one person to another is dependent on attitude as well as physical presence. Between units, communication is also dependent on equipment capability, and it is the responsibility of the leader to make sure the receiving communications equipment and personnel are up to standard. In both cases, the motivation of the person receiving the message must be considered. It is obvious that if the receiving individual does not understand the language of transmission, the message will not get through, and it is just as true that the message will not be received if the individual does not want to receive it. Admiral McKee notes that

counseling on fitness reports/evaluations is probably the best example that I can think of to illustrate the need for an officer to properly communicate that he or she has a subordinate's best interest at heart. This includes commending on strong points as well as sometimes getting an unwelcomed message across on some area of deficiency. While it is generally easy for people to accept praise, the reverse may be said for people when they hear of their shortcomings. Since most of us have delusions of our approaching perfection, it can be an enlightening moment to hear of any discrepancies in performance. It's therefore important that the individual being counseled be receptive to any discussion on this point. It is easy to listen to praise or to dispense it, but a productive discussion of below-par performance is some-

times difficult. It is the responsibility of officers to counsel and assist subordinates, and it must be done with skill and understanding to be effective. The best interest of the individual and the overall unit will then best be served.

A primary consideration in the communications process is to ensure that the individual not only is ready to receive the message but wants to do so, as well. This desire to receive a message depends partly on the receiver understanding both the importance of the message and the effect it can have on the unit. The receiver in this instance is the ultimate receiver, not the person or persons who transmit a message between sender and receiver. Obviously, not everyone in the transmission process will understand the importance of all the messages they send. The receiving commanding officer, therefore, must ensure that training and experience and equipment capability make it possible for the sender's message to be received on board.

Certainly it can be assumed that well-trained, disciplined, and loyal officers will want to receive another officer's message, but the sender should keep in mind that the receiver has many competing matters to consider. Thus, messages sent in a way that indicates to the receiver the value of receiving the entire message are more likely to get the receiver's attention. The leader must also be on the lookout to avoid words with ambiguous meanings. Some words may have many meanings, and it is best to avoid these unless it is perfectly clear what is meant when one is used. Semantics is the study of words and word usage and may help the officer in avoiding this problem. Before messages are sent between people, the people usually have established a relationship by personal contact or by reputation. Preexisting conditions such as this affect communications, and thus interfaces between groups and individuals set the stage for future communications between these same groups: communications generally are transmitted well between "friends." By communication style and personal contact, then, a unit or an individual develops a reputation of generally communicating something that is worthwhile to listen to. Receivers of a message from these individuals or units will want to listen, for they realize that they will benefit from what they hear.

This puts the onus on the sender to ensure that the message being sent is, in fact, important to the receiver and will be of help and value. It is necessary, then, for the sender to understand the mission and goals of the receiver; only then can he or she be assured that the messages being sent meet the needs of the receiver and will not become just more paperwork for the receiver to process. There are five principles to help increase the accuracy of the message:

1. Ensure that the message is relevant to the receiver.

2. Reduce the message to the simplest possible terms.

3. Organize the message.

4. Repeat key points (verbal messages).

5. Focus on essential aspects only.

It is useful for individuals and organizations to review messages with those who sent them, whether through data links or in person, and to discuss the value of the messages sent and received. If an individual receives a message that does not seem to have any practical application to the matter at hand, then he or she should raise a question with the sender as to the importance of what was sent, so that the receiver can discover what the importance is. There is a dual responsibility in this relationship: not only is the receiver required to check with the sender if he or she does not understand the reason for the message, but the sender must double-check from time to time to make sure the receiver is getting the message that was intended.

The main point here is that messages are sent for a purpose: to convey information to others which may affect their actions. Messages will be better received if the receiver sees a value in the message (this is a responsibility of the sender), and the action of sending a message is only completed if the receiver receives the message and the sender knows that the receiver understands the meaning the message was intended to convey.

Chapter 6. Study Guide Questions

1. What is a primary consideration in the communication process?

2. What are five principles that help to increase the accuracy of a message?

3. What is the dual responsibility in effective communication?

Vocabulary

commend
deficiency

semantics
onus

Chapter 7.   Setting the Example: One Form of Communicating

Consider the following clichés:

"Actions speak louder than words."

"I'm from Missouri. Show me."

"Do what I say, not what I do."

We tend to toss off these well-worn phrases, and we tend to answer them with a "Ha!" Perhaps because they are so often repeated, they have lost credibility, especially in today's verbal-video world, where "image" or reputation is frequently made in a few paragraphs or in 90 seconds on television. Disregarding the message behind these clichés is a trap into which naval officers should not and cannot fall if they expect to be effective in the hard, real world of the U.S. Navy or Marine Corps. One of the foremost responsibilities of a leader is to serve as the example or model for followers in everything that he or she does or is. The military expects that its officers will be an example to their troops in the performance of duty, in the sharing of hardships and danger with the troops, and above all, in the high standards of moral and ethical behavior.

This is not to say that verbal and written communication is not important. On the contrary, the talent to communicate effectively with language has been the mark of great leaders throughout history.

The words of Winston Churchill and of Franklin Roosevelt during World War II were a great inspiration to the people of their nations. Douglas MacArthur, Chester Nimitz, and William Halsey also knew how to use the language to great effect. But all of those leaders had something else in common: they acted on what they said. They backed up their words with behavior and performance that made the words they spoke and wrote true and therefore meaningful.

One key to effective communication in leadership, then, is being articulate—and that does not necessarily mean "fancy"—and matching articulateness with action. When actions and words are not combined effectively, they are out of synchronization. Everyone has met a great "embellisher," a person who has more big-fish stories than anyone else but who never seems to produce the fish—or even the fishing pole. The embellisher is wonderful at happy hours but not very good during working hours. That person is not someone with whom many would choose to go to war.

Then there is the person who says nothing, forcing everyone else to interpret his or her actions. It is not always bad to say nothing. However, actions in this complicated world are not always simple or clearcut. They sometimes leave people saying to themselves, "What's going on?" Everyone has been perplexed by the unexplained actions of another at least once. A leader is above all a teacher and, like all teachers, constantly under the very acute, and not always charitable, observation of pupils. They will, perhaps instinctively, model themselves on their leaders—even to the extent of copying their idiosyncrasies.

So there is a simple formula: *Say it well* plus *do it well* equals *successful communications*. Right? Almost, but not quite. There are two intervening variables that make a big difference. First are the expectations of the officer's audience, whether it is made up of peers, crew members, or the general public. Second is the stereotype within which the officer operates (everyone operates within a stereotype). These two factors are separate, but they act together in the fine art of human communications.

An individual carries the stereotypes belonging to the career and lifestyle chosen as well as to the organizations joined. These categorize this individual. Corporate presidents have different images or stereotypes than rock stars or professional football linebackers. People conjure up a mental image of an individual well before they meet him or her. Therefore, they expect that person to display a certain pattern of behavior and speech. These expectations are not always valid, but they are there. Admiral Taylor:

Actions speak louder than words is the expression, and I think that's true, but I don't think actions alone can do it, and I don't think words alone can do it. My style has been to keep the troops well informed. When you're in command of an aircraft carrier with 5,000 guys scattered out in 2,000 spaces, what keeps it all together from my experience is a commanding officer who gets out there on a regular basis for a few minutes. If he gets out there and he talks for too long, then he's losing the crowd, but if he boils it down to the right kinds of things, then people are going to listen.

You also have to get out among the troops, and you have to be true to your word, to what you tell them, and you have to set an example of leadership. You have to be up early, and up late, and work hard in the middle. And if you don't care about what time you get in there, what time you go home, if you don't have the philosophy of first in, last out, and I'm going to work hard in the middle, then you're going to get left by others, because that is the old work ethic. The leader has to set the example; that is something that junior officers must fully understand. There are an awful lot of people who know they're smart and sharp, and then they'll get whipped by the turtle

who gets up early and stays up late. So your actions and what you do and making good on your word and setting a good example in every respect, of being personable, keeping the old positive attitude out there are important.

Let me relate an experiment I ran on *Coral Sea* with respect to actions and positive attitude. After a bad night with three or four hours of intermittent sleep, 20 phone calls, as we made our way through a lot of busy water, I finally got up at 5:30 and started walking on the bridge at ten minutes of six. As I walked through the pilot house, one of the crew said, "Good morning, Captain." I replied, "Good morning, troops. How're we doing here? It looks like it's been in good hands tonight. You guys are all looking smiley faced here, I guess everything is in good shape, not a worry, right?" and I walked out on the bridge and I saw the OD and told him, "It's a great-looking day, and I think we're going to have a good day today." I set a tone of positive and good and everything's under control and smiley face, and all those guys on the sound-powered hook-ups, they're talking down into the ship, saying, "Look, Captain's on the bridge, everything's going good." The tone is set. A positive tone.

But then on a couple other nights when there were two or three hours' intermittent sleep, maybe I got a bad letter from home, bills were piling up, and the executive officer was not hacking it, and I walked on the bridge and they said, "Good morning, Captain," and I said, "Same to you, fellow." You put on a little old glum face and the first thing you say to the OD is "Get the XO up here," then you get the XO out on the wing of the bridge and the guys are saying, "Hey, the skipper's up here chewing on the XO," and then all of a sudden there's another tone going out through the ship. But what you've done is squash enthusiasm.

As a leader you have to be able to absorb that shock, just take it and take it and take it. It's the greatest thing in the world, it makes you stronger and better for it. It's a great privilege and honor to lead troops, but it's a price you pay. You have to be able to set a tone and keep it there every day, and you do that by your actions, just little subtle ones, but you have to have them, and if you're going to want the troops to be positive, you've got to be positive. Case in point, the 1980 presidential election, we have a leader with all his training who gets up and says everything's okay. We have another guy up there saying everything's terrible. Now, political philosophies aside, to me it said who do we want to lead, the guy with the smile who says everything's going to be okay, trust me, or the guy who says, hey, we're in a whole lot of trouble? Everybody's always in a lot of trouble. But your leadership has got to get in and say we're going to whip this problem, stay with it. The same thing is true at a breakfast table in the families of our country. If the mother and father are coming down to breakfast and there is a positive tone, then it goes out. And then achievement, you check back in at dinner time, and everyone has a positive glow there. But turn it around the other way. If the bills are piling up and there are internal problems in the family, you can squelch real enthusiasm for going out and whipping the world by just the tone that gets set at breakfast.

Naval officers who are graduates of the Naval Academy, as well as Academy midshipmen, are stereotyped immediately. The image or reputation of the Naval Academy follows its students and graduates, and its students and graduates reinforce or erode that image. The process is synergistic. The people naval officers interact with have expectations of them based on the stereotype of Naval Academy graduates. In order to communicate effectively, naval officers must understand their own stereotype, because what they say and do will be interpreted within their audiences' expectations, at least at first. The leader sets the moral tone for subordinates by the example of integrity he or she provides in both official duties and private actions. Honesty cannot be instilled by contract—but may be enhanced by education about its importance to mission accomplishment and by example.

This does not mean that an individual must be or live the stereotype. But the individual must understand the stereotype in order to carve his or her individuality from it, to redefine or enlarge the stereotype with a personal image. In the process, the individual is able, ultimately, to communicate from a platform of his or her own making. The process begins, however, with an understanding of the expectations and the stereotypes that obtain.

One point must be made here. As each person redefines the stereotype with an audience, there is a responsibility not to tarnish, but rather to enhance, the fundamental stereotype within which the person has chosen to operate (e.g., the Naval Service). Fulfilling that responsibility effectively brings individual benefits and institutional rewards.

How does this process work in practice? Lt. Bobby Goodman, USN, was a bombardier/navigator in an A-6 shot down by groundfire over Lebanon in 1983. His pilot was killed. Lieutenant Goodman was captured, taken to Syria, and held there for more than a month. He was released and returned to the United States with Reverend Jesse Jackson to a hero's welcome at Andrews Air Force Base. He was subsequently sought after by all of the television talk shows and other media from around the world. Heady stuff for a young officer to handle all at once, but he did exceptionally well, and when he returned, what he had to say about himself, the mission, and his captivity reflected a cool, deliberate, intelligent, and professional manner. What he said and how he said it matched his actions before, during, and after his captivity. What is more, all of these things fulfilled the expectations people have of a professional naval aviator who graduated from the Naval Academy. At the same time that Lieutenant Goodman carved for himself an individuality, he brought credit to himself and to his service.

Finally, when officers communicate verbally or nonverbally, they must be concerned with context. When, where, and how they say or do something makes a great difference in the way it is interpreted. If someone yells "Beat Army" at a rally in front of Bancroft Hall, that is very positive. What if someone yells the same thing with the same intensity at the Tomb of the Unknowns on Memorial Day? Same words, same volume, but a different message that generates a different image for the yeller. A ludicrous example? Perhaps. But the point is clear. Be aware of context. Be aware, too, that context is dynamic. As society and culture change, as events change, opinions and mores change style, so context changes. Consequently, no one can communicate well without being aware of the surroundings. Army Chief of Staff Gen. Edward C. Meyer put it well in 1980 when he said, "The obligation of service and commitment inherent in the military ethic imposes burdens not customary in the larger society, where obligations are normally contractual in nature and limited in degree of personal sacrifice expected. For the soldier, the obligation is complete: to the death if necessary."

Remember, then: Say it well. Do it well. Know your stereotype and yourself and your audience's expectations. Be aware of the context. Stay in synchronization or, in today's cliché, "Be a together person."

MCPON Sanders: *The crew must know your policies, not necessarily by you repeating them but by personally setting the standard, yourself, as well as requiring the crew to meet standards. It's really a matter of personal example in everything you do and how you approach your work. If you set a good example, others will follow.*

Chapter 7. Study Guide Questions

1. What is one of the foremost responsibilities of a leader?

2. What do great leaders like Winston Churchill, Franklin Roosevelt, Douglas MacArthur, Chester Nimitz, and William Halsey have in common?

3. What is the simple formula for successful communication?

4. What stereotypes do individuals carry with them?

5. How does the leader set the moral tone for subordinates?

Vocabulary

credibility

cliché

synergistic

articulate

Chapter 8. Communications as a Component of Planning

In the process of developing plans for the future, the officer is afforded the opportunity to communicate both formally and informally. To ensure the development of useful plans, the officer must engage in a continual exchange of ideas, requirements and objectives with others. In fact, the process of planning is sometimes more valuable than the plan itself, and can be viewed as a communications exercise. Planning requires a thorough and common understanding of the organization's mission and function, and it formalizes the steps that will be taken to execute them in the future. Each word or group of words has different meanings and connotations, and does not convey the same ideas to every person; thus, special care must be taken in communications.

The process of establishing organizational goals provides another opportunity for everyone concerned to discuss where the organization is going. Normally this process entails setting up long-term goals or objectives; a set of more specific, short-term objectives is established periodically. The establishment of objectives is a top-down evolution, with each organizational layer implementing the senior's objectives as they apply. Although an active exchange will result in understandable and achievable objectives, the leader ultimately sets the objectives. Again, the communication leading up to the actual setting of objectives is valuable.

As Admiral Zumwalt points out,

the importance of communication is (a) to make sure that what is wanted is understood, (b) to make sure that the understanding includes why the order or assignment makes sense, and (c) to make sure that the sense includes an explanation of how the specific assignment fits within the objective of the day or the month or the year. These three aspects of communications are sufficient to convey an order, but the order is often followed better if in addition the officer can make it part of the fun of the day or the month or the year.

The officer who has first-hand contact with his

people will be more successful than one who does not, because they will be familiar with the officer's style, and the officer will be aware of their capabilities. A commanding officer, for example, might get around daily to see what is going on in various parts of the ship, to drop into chief's quarters for that cup of coffee while he listens carefully amidst the banter to pick up where there is a problem that is not otherwise surfacing. He can then go back and ask a question through the chain of command that will lead to the necessary insights and changes. The successful leader analyzes the situation when something has not gone right, and determines whether or not in part or in whole it was the result of not communicating properly. If he concludes that that is the reason for the failure, then he should seek to improve his or her subordinates' ability to communicate.

Oftentimes the senior (rating officer) and the subordinate can use organizational objectives to establish individual objectives. The communication surrounding this procedure defines the future relationship of the two and determines the type of work the subordinate will do over the next rating period.

Following up on plans and objectives established for the future is a necessary part of discipline. The follow-up should be simple and avoid excessive paperwork. The follow-up process must be well understood by both the senior and the junior, and it must permit the junior to ask questions concerning his or her progress. Without a followup system, human nature will lead individuals to expend all of their effort on immediate issues or crises and neglect the direction of the organization. A means of measuring progress toward goals and objectives, or periodic plan assessments, is essential.

In some instances it is beneficial to formulate a method of measuring progress at the time a plan or program is approved. This technique avoids misunderstanding and allows future success or failure to be easily determined.

On the general subject of communication, Admiral Holloway makes the following point:

when the officer deems the delivery and comprehension of a message to be very important, he should not hesitate to ask for an acknowledgment of the message. For example, if an air raid on a city was scheduled, but the commander received word from the White House not to deploy because a U.S. government envoy was arriving in the city that day for a peace conference, the commander's message to the front would probably read: "Cancel air operations tomorrow in Vietnam between 1000 and 1800 local time. Repeat, cancel scheduled operations. Acknowledge this message." Because it would be essential that no one missed that instruction—and end up sending a combat flight in there when the envoy arrived—the commander would request acknowledgment.

On the other hand, many messages do not require an acknowledgment. The person who sends the messages can tell by the messages he receives within the next week or so whether the original message was understood.

If he considers it necessary, the officer can always ask his senior to acknowledge a message. Say the leader of a unit is told to deliver an intelligence officer to a destroyer on the gun line, and it is very important that the intelligence officer get up there because he has some information. If the weather was bad, the unit leader would immediately send the word back: "Unable to deliver Lieutenant Jones to USS *Treadway* because of bad weather. Repeat, unable to deliver. Confirmation requested." The leader must be sure that the commander receives the message; if he does not receive it, he will assume that Lieutenant Jones delivered his very important information, and he will make his plans accordingly.

Chapter 8. Study Guide Questions

1. What should the leader do to ensure development of useful plans?
2. What does Admiral Zumwalt point out about the importance of communication?
3. Why is a follow-up system on plans and objectives important?
4. According to Admiral Holloway, what should be done if the delivery and comprehension of a message are very important?

Vocabulary

insight
connotation

Chapter 9. Written Communications

Much of the routine, day-to-day operation of the Naval Service is governed by written orders. U.S. Navy Regulations, OPNAVINST 3120.32A (the Navy SORM), Marine Corps Orders, and Uniform Regulations are examples of written orders that govern the way in which naval officers go about their business. Of particular importance is the detail these written orders contain and the frequency with which officers benefit from the guidance they provide. They deal with relatively complex, multifaceted issues and generally are not subject to rapid change or modification based on the circumstances surrounding their implementation.

The written instructions and notices published by a command or a higher headquarters usually fall into a similar category, but the scope of their coverage is less broad. Similarly, it is usually advisable to utilize written orders when citing or quoting a higher authority, thus eliminating the possibility of error or confusion through misinterpretation.

Admiral McKee states that

in general, written orders are for more complex tasks or tasks requiring detailed explanations. Standing orders, or orders that you want to place in a book of information or on bulletin boards, or things that you want to have or use for some time should also be written.

Admiral Larson:

Written orders are very important when the orders are complicated, when they are extremely formal, when they are very important to the safety of the ship. Now one example here are your captain's standing orders to the officer of the deck, or your captain's night orders. You don't want to rely on word of mouth passing from person to person, but you want to have certain basic principles in writing so they can be referred to, and referred to frequently. Written orders convey a sense of formality and provide a common frame of reference for all of your actions and all of your decisions. I think we all realize how easily the verbal word can be confused once it's passed from person to person to person. We've all played that old parlor game where you whisper around the table in somebody's ear and you don't recognize the result that comes after five or six people have passed a sentence. So I think that all the following should be written so you have that very important frame of reference: (1) very formal things that are the basis of your organization, (2) complicated orders, (3) things that will need to be referred back to and spot-checked because you want to make sure that you're doing it exactly right, (4) things in highly sensitive areas where you need to be sure, such as in rules of engagement or ships' actions, and (5) things that are critical to a ship's success. Verbal orders can be given to augment that foundation as you operate your command.

Operations Orders (OPORDs) written to govern the operations of a fleet or the conduct of a fleet exercise are documents that, because of the complexity of the evolutions they control, must spell out in detail what procedures are to be followed and how operational plans will be executed. Some sections of an OPORD, such as a schedule of events, are subject to change as an activity evolves, but the basic governing document normally remains in effect throughout the duration of the activity.

From these examples of written orders emerge several common features. The following characteristics are normally associated with the use of written orders in the Navy:

1. The topic addressed by written orders usually is complex, is broad in scope, and requires a considerable amount of detailed explanation to be conveyed effectively.

2. Implementing changes to written orders requires time.

3. The guidance provided by written orders is usually of an enduring nature, governing an activity for a considerable period.

4. Personnel subject to a written order are strictly bound to adhere to its provisions; written orders leave little or no room for interpretation or personal preference.

The use of written orders is not restricted to matters of policy and broad interest, however. There may be occasions when it is in the best interests of a senior to provide written guidance to a junior. For example, a complex task, a need for strict accountability, or a need to correct noted performance deficiencies would be better handled with a written, rather than a verbal, order. The specific responsibilities of certain watchstanders, such as sentries or sounding and security watches, should be clearly spelled out in writing.

The provision of policy and guidance through a written order is typified in shipboard orders such as the commanding officer's standing orders and the engineer's night orders. A junior officer can apply the concepts delineated in these orders when writing his or her own orders for a unit as small as a division. In these orders, the junior officer can specify divisional policies, work and cleaning assignments, and operating procedures for the personnel concerned.

The concepts in the following list can help an officer determine whether written orders can be used effectively and efficiently to accomplish a specific mission.

- Do use a written order to promulgate policies or assignments that have wide application.
- Do use a written order to cover complicated issues, particularly when subordinates may need to refer back to the order for details or additional guidance.
- Do use a written order when there is any indication that a subordinate might deviate, deliberately or otherwise, from verbal guidance and must be held strictly accountable.
- Do use a written order when it is anticipated that its provisions will remain in effect for some time and that sufficient lead time will be available to issue any necessary changes to the order.
- Do use a written order when continuity is needed and a succession of persons is involved.
- Do use a written order when cooperation or coordination between several organizations, units, or work groups is required.
- Do use a written order when it must be issued to persons or activities in widely separated locations.
- Don't use a written order when timeliness is paramount.
- Don't use a written order for very simple instructions.
- Don't use a written order when rapid changes in tasking or circumstances are expected.
- Don't use a written order as a substitute for leadership and instruction.

Admiral Long suggests that

it is very important for ships or squadrons or base operations to have written instructions, particularly those dealing with the operation of complex equipment. That equipment must be operated properly and maintained properly, and this includes such things as preventive maintenance schedules, operating procedures, safety precautions. There must also be written procedures and doctrines for essential personnel matters, such as how someone gets promoted or what someone's benefits are. There should be written guidelines for the necessary reports that are written and the inspections that are conducted, too.

Any organization needs to be careful, however, to make sure that written procedures, paper work, and check-off lists serve primarily as a guide or a reminder for most people. They should not become so indispensable that people cannot function without them. If people fail to follow the procedures, it will be harmful to their equipment or to themselves or to their own welfare or to morale, and people must understand this. But it is impossible to cover every situation, so it is essential that people try to adapt general principles to specific situations. An old tongue-in-cheek rule is: As a last resort, let's shift to common sense.

As an officer advances in his rank he will no doubt rely more and more on informal discussion. But when a senior command is communicating with a junior command, it is incumbent upon the senior command, regardless of the amount of informal staff discussion that takes place on a significant issue, to put their views in writing rather than passing it up and down the line by staff to staff on a telephone.

A leader must know what is going on, and clearly one way he can find out what is going on is to receive written and oral reports through a formal chain of command. An officer can also keep informed by inspecting the spaces or "walking the ship." Another very important way is to talk to his people.

MCPON Sanders: *I would recommend to the junior officers that they use written orders when (1) the orders of operation are important and need to be followed exactly, (2) you desire to hold an individual strictly accountable, and (3) there is complicated work or a lot of things to get done within a short period of time. In the last case it is also important to set a priority listing of what you want done by order of importance.*

Chapter 9. Study Guide Questions

1. When is it better to communicate in writing rather than verbally?
2. What is a key advantage of written orders, according to Admiral Larson?
3. What characteristics are normally associated with the use of written orders in the Navy?
4. When should a written order be used, and when should it not be?

Vocabulary

multifaceted   accountability

Chapter 10. Oral Communications

And ye shall compass the city, all ye men of war, and go round about the city once. Thus shalt thou do six days . . . and when ye hear the sound of the trumpet, all the people shall shout a great shout; and the walls of the city shall fall down flat. (Joshua 6:2–5)

These are some of the earliest recorded oral military orders. According to the Bible, God spoke these orders to Joshua and directed the fall of Jericho. This biblical passage provides a good introduction to a study of oral orders.

An officer can expect to give and receive thousands of oral orders during his or her career. An order, according to the *Joint Chiefs of Staff Dictionary*, is "a communication, written, oral, or by signal, which conveys instructions from a superior to a subordinate." The situation dictates the manner in which this communication is made, but as long as the issuer of the order possesses the requisite authority to issue the order, any manner he or she deems appropriate is acceptable.

Admiral Long suggests the following:

in communicating orally, leaders should adjust their style and their content to fit their audience. People who communicate in a monologue in most cases are not communicating as well as they could. Officers need to develop the ability, whether in a social environment or in a professional environment, to listen to people.

It is especially important for officers to give subordinates an opportunity to express their views. Not only does the senior gather more information this way, but he also has subordinates who become a part of the organization and a part of the business of running such an organization because they are able to express their own views and know the senior is listening. Of course the responsibility still remains with the senior, but he will benefit if he can generate within his subordinates not only candor but also loyalty to carry out his decisions after they are made.

It is impossible to have written orders on every aspect of professional life, and this is particularly true in an operational capacity, such as aboard ship or in an aircraft or in an organization of Marines. There used to be a saying within the submarine force, which probably has as many written directives and instructions and procedures as any military organization, "We do not have an instruction forbidding elephants down in the reactor compartment." This was just a way of saying all situations cannot be covered, and at some point those in command will have to make on-the-spot decisions.

Good orders meet three criteria. First, they point the way to a goal. To achieve this goal, the order describes a well-thought-out plan and specifically tasks each person with a job that contributes to the completion of that plan. The heart of an order is the plan it describes. A good order also clearly describes the role every person will play in the plan to achieve the goal. Furthermore, only orders that you intend to enforce should be instituted. Orders that are issued just for the record are useless.

Second, orders have to be understood by the individuals to whom they are given. Helmuth von Moltke, in the 1800s, said, "An order that can be misunderstood will be misunderstood." The heroic charge of the Light Brigade, made famous by Alfred, Lord Tennyson in his poem of the same name, was a military blunder caused by misunderstood orders. The 600 brave cavalrymen who rode to their deaths were carrying out an order from their commander to "Attack the guns!" The guns that the commander wanted them to attack were a small battery off to the side of the valley. Unbeknown to the commander, this battery was hidden from the brigade's view by a berm. The 600 men saw only the main battery of guns at the far end of the valley, presumed these were the guns they had been ordered to attack, and rode to their deaths, dutifully carrying out misunderstood orders.

Rarely is an order of such consequence given. As officer of the deck on a destroyer, however, an officer has responsibility for the safety of the entire crew, and he or she must word orders carefully. In an emergency, the helmsman who is issued an irregular order may misunderstand it, with disastrous consequences. The maintenance officer of a helicopter squadron who gives an order to "replace a worn-out linkage" on a helicopter could cause the death of a crew six months later; the accident investigation will show that the wrong linkage was replaced. The decisions an officer makes and the orders given will, by the very nature of the job, affect the lives and safety of fellow officers and the enlisted personnel. Orders must not be misunderstood.

The third part of a good order is timeliness. Personnel need enough time to carry out orders in accordance with the plan. Military evolutions are often complex, requiring long set-up times and coordination with other units. An order to launch a noon air strike given at 1145 is impossible to carry out, and if this strike is in support of a noon amphibious landing on the beach, the consequences, again, could be disastrous.

These are the three essential criteria of a good order. It clearly tasks each person with a job in accordance with a plan, it is easily understood, and it is timely. Orders may be in written, oral, or signal form, but oral orders are the oldest and by far the most common form of orders given. Officers will learn through experience when it is best to use oral orders, and they will discover the problems peculiar to their use, but the following are generally accepted guidelines for issuing oral orders. Four situations are particularly well suited for using oral orders: when there is a need to clarify written orders and provide minor details; during an emergency; during an ongoing operation; and when an officer is working with a thoroughly trained team.

Adm. Horatio Nelson, perhaps the greatest naval tactician who ever lived, was famous for his oral orders. He would gather his ship captains together before a battle and explain his plans and expectations. After these verbal briefings, written orders would be issued and the whole plan would be started by a simple numerical sign. The prebattle oral communication was the essence of the "Nelson Touch." It ensured a clear understanding of the written orders and gave Nelson's captains an opportunity to ask questions and clarify the details of the battle plan. Oral orders are still an excellent tool for clarifying written orders and for providing minor details to personnel.

Emergency situations often require verbal orders. Admiral McKee comments: "Verbal orders are issued when you don't have a great deal of time or when the individual is very knowledgeable about the task and it takes very few words to delineate a task to that person and you think that they'll understand what you say without detailed explanation." In August 1814, President James Madison and several members of his cabinet observed the beginning of the battle of Bladensburg. Secretary of the Navy William Jones mentioned to the president that if the British were to win the battle and march on Washington, D.C., ships and stores at the Washington Navy Yard would fall into enemy hands. Jones recommended their destruction should the British advance. Madison agreed. Later that same day, Jones visited Capt. Thomas Tingey, commandant of the yard, and issued verbal orders to burn whatever the British could make use of should they reach the city. That night Tingey was forced to follow the secretary's command.

During an ongoing operation, timeliness is an important factor. Oral orders allow an officer to give instructions to personnel quickly and without interrupting their work. A captain's verbal order to "Fire" in the midst of battle, or rudder orders to the helmsman, could not effectively be written down and delivered. Oral orders ensure timely communication and minimize the interruption of the work at hand. They are an excellent way to give direction to personnel during an ongoing operation.

Lastly, verbal orders are well suited for directing a thoroughly trained team. Experienced boatswains know what they and their teams should do when they hear orders to "single up the lines." A thoroughly trained crew knows where to go and what to do when general quarters sound over the 1MC. Thorough training beforehand enhances the communication and understanding between senior and subordinate and permits the effective use of verbal orders in virtually any situation.

General Rice notes that

time is one factor to consider in determining whether orders should be written or spoken. It is far better in terms of staff coordination to have written orders. A staff evaluation, for example, should be written, because the people have something they can turn to. In emergency situations, fast-breaking situations, verbal orders are called for. The key to successful verbal orders is for people to know the officer, to have worked with him and trained with him. If the officer and the troops have done things over and over again, have practiced different things, and the troops know that the officer can save time by giving a verbal order, they will respond to it quickly and well.

Problems associated with verbal orders stem from the same qualities that make them so well suited to the preceding situations. The speed and ease with which oral orders can be given make it easy for a commander to issue bad orders. A hastily given order can be misunderstood because it provides only partial or inaccurate information. Or a commander, under pressure to deliver an order quickly, could decide on a bad plan—for which there can be no good orders. Subordinates will know and measure their officers most by orally delivered orders. Officers must be careful not to let the ease with which they can deliver an order orally diminish the care they take in leading subordinates.

As a subordinate to seniors, an officer should recognize that the action taken on verbal orders, should it lead to debacle, leaves him or her open to scapegoating. Safety cannot be found, however, by demanding that everything be in writing, but only by demanding that both seniors and subordinates command respect and display trust through their professional behavior. Without respect and trust up and down the chain of command, an oral communication system cannot function effectively. In his memoirs, Viscount Montgomery of Alamein wrote: "A commander must train his subordinate commanders, and his own staff, to work and act on verbal orders. Those who cannot be trusted to act on clear and concise verbal orders, but want everything in writing, are useless."

Chapter 10. Study Guide Questions

1. What is an order?
2. What three criteria must good orders meet?
3. What are four situations particularly well suited for using oral orders?

Vocabulary

order	debacle
requisite authority	scapegoat
blunder	

Chapter 11. Avoiding Communication Pitfalls

Orders are absolutely necessary; they tell individuals how to act in their formal positions. When giving an order to subordinates, the leader must use a straightforward approach and the simplest of terms to convey what he or she wants them to accomplish. Orders should leave no room for interpretation and should contain no words with a hidden or "double" meaning. Too often the word *should* is used in orders when the officer really means *must* or *will*. Before issuing an order, an officer must always consider whether or not the order is explicit.

As Admiral Holloway points out,

orders are most likely to be misunderstood either when the senior who drafts them does not take the time or care to make them clear, concise, and unambiguous, or when a subordinate is unable, for some reason, to understand what he is being told. Very often an order is unclear because an officer relies on his aide or assistant to convey his wishes. The officer who writes his own orders can be sure they are worded so they can be understood. An officer should instruct his staff to check with him if an order is unclear, because sometimes assistants are reluctant to disturb their boss for that purpose. On the receiving end, subordinates sometimes have difficulty understanding the intentions of their leaders, so officers must eliminate ambiguity from their messages, both written and oral.

If an explanation or background information is necessary for the proper execution of an order, and time permits, the officer should provide this information prior to issuing the order. Then he or she should issue the order as a checklist. Sea stories or superfluous materials are not required, and they will only create diversions and confusion for the crew or the troops. An order should not be entwined with an explanation or overview.

People work better when they understand the purpose of their efforts, so a leader can often achieve better results by providing subordinates with an overview. Giving subordinates sufficient background material will make them feel that they are involved and are taking an active part in the exercise or work effort. Subordinates who understand the reasons for an assignment are also better able to exercise their judgment and make adjustments as the situation changes or as problems arise. Officers who keep everything to themselves and practice the "mushroom" theory with subordinates only create problems. A leader cannot keep people in the dark and expect them to function effectively and to their full potential or to feel that they are a part of the "team."

Leaders must know what they are going to say. Unorganized thoughts result in an unorganized effort, so the leader should consider whether the ideas are clear and concise and, if there is time, should bounce them off a roommate or colleague before expressing them to subordinates.

From the top on down, orders are sometimes given without thought to priority of accomplishment, redundancy, or consistency. Feedback and rapport up and down the entire chain of command are required if orders of this nature are to be avoided. Following the chain of command may act to slow down transmission. Despite the dangers of such a delay, important communications should nevertheless pass through every level of the chain of command, from top to bottom. Subordinates in the chain must continually discuss problems and priorities among themselves and with their commanding officer. Rapport among people will preclude their issuing conflicting orders and will permit an effective and coordinated effort.

Before issuing routine orders, a leader should discuss priorities with others. He or she must keep in mind that the desires of the commanding officer receive top priority, even though the commanding officer may not have made the task a top-priority item.

Leaders must not take the drive out of subordinates by chipping away at their egos. A leader has to avoid threatening the status or ego of a subordinate at all costs, and he or she must recognize a situation where this is likely to happen before getting into the middle of it.

Giving false expectations, such as early liberty, a day off, or basket leave, for the accomplishment of a task that cannot be accomplished within the time frame or is beyond the subordinates' capabilities is damaging to the morale and cohesiveness of the unit. Goals have to be attainable, and a consistent policy, such as permitting the chief petty officer or staff noncommissioned officer to control leave or early liberty based on job accomplishment, promotes high morale within the division or unit. Work that has been done correctly and in a timely manner can always be rewarded with special privileges. This tends to keep the unit in a high state of readiness.

To avoid having orders distorted as they pass through the chain of command, a leader should make sure the leading petty officers or noncommissioned officers use wheelbooks or notebooks at quarters. After quarters, the leader can observe and listen to the LPO/NCO putting out the same information the leader just provided. If there is a distortion or other problem, the leader should call the LPO/NCO aside and settle the matter as soon as possible. Many administrative failures can be traced to seemingly discretionary orders to which absolutely strict compliance was expected.

A leader must establish a rapport with subordinates that permits them to ask questions without fear of receiving a sarcastic reply or a foul look. The leader must provide the LPO/NCO the opportunity to ask questions, either during a meeting or afterward in private.

If the leader must criticize or correct someone, it should be done in private, and the leader should always finish up the meeting by reminding the individual of some of his or her good points and recent accomplishments. Leaders have to treat subordinates as competent individuals until they prove otherwise. Additionally, leaders must respect someone who is honest and sincere. Everyone wants to do well and to please the senior, and the senior should keep that in mind and treat subordinates as he or she would like to be treated.

1. According to Admiral Holloway, when are orders most likely to be misunderstood?

2. What is accomplished by giving subordinates sufficient background material?

3. What should a leader avoid doing to a subordinate at all costs?

Vocabulary

unambiguous
straightforward

Appendix: Partial List of Contributors

THE UNITED STATES

Chiefs of Naval Operations
Adm. Arleigh A. Burke
Adm. Robert B. Carney
Adm. Thomas B. Hayward
Adm. James L. Holloway
Adm. David L. McDonald
Adm. Thomas H. Moorer
Adm. Carlisle A. H. Trost
Adm. Elmo R. Zumwalt, Jr.

Commandants of the Marine Corps
Gen. Robert H. Barrow
Gen. Wallace M. Greene, Jr.
Gen. P. X. Kelley
Gen. Lemuel C. Shepherd, Jr.
Gen. Louis H. Wilson

Other Contributors
Sgt. Maj. John L. Horton, USMC
Adm. Robert Long, VCNO &
    USCINCPAC
Prof. Karel Montor, U.S. Naval
    Academy
Vice Adm. Hyman G. Rickover,
    Director, Nuclear Power
    Division, BuShips
MCPON Billy C. Sanders, USN
    (Ret.)

ENGLAND

Adm. Sir Nicholas Hunt,
    GCB, LVO, Royal Navy,
    CINCFLEET

FRANCE

Adm. Paul de Bigault de
    Cazanove, Commandant Chief
    of the Atlantic Theater

WEST GERMANY

Adm. Karl Clausen, Commander
    North Sea Forces
Adm. Friedrich Ruge, CNO

JAPAN

Chiefs of Naval Operations
Adm. Takaichi Itaya
Adm. Teiji Nakamura
Adm. Kazuomi Uchida

Glossary

ACCOUNTABILITY—the responsibility and obligation imposed by law on military personnel for their management of personnel, property, or funds assigned to them.

AMBIGUITY—something susceptible to multiple interpretation.

ANIMATE—to activate, give life to, energize, or arouse.

APATHY—a lack of emotion or feeling.

ARTICULATE—to speak clearly, enunciate, or verbalize.

ASSESSMENT—an estimate or evaluation of an object or event.

ATTENUATION FACTOR—the ratio by which something is diluted or lessened in value, amount, or intensity.

AUTOCRATIC—pertaining to rule by a person having absolute or unrestricted power.

BANALITY—the repetition of a worn-out idea; something drearily predictable.

BLUNDER—a stupid or grave mistake.

CAMARADERIE—good will and rapport between or among friends.

CANDOR—frankness of expression; impartiality.

CARICATURE—a drawing or other representation of a person or thing that exaggerates its distinguishing features for satirical effect.

CENSURE—to express blame or disapproval.

CHAIN OF COMMAND—the organization of authority from highest to lowest in a military unit—the succession of people through which command is exercised.

CLICHÉ—a trite or overused expression or idea.

CLINICAL—based on or concerned with the actual treatment of patients rather than research or theory.

COERCIVE—tending to force to act or think in a given manner; compelling by pressure or threat.

COGNITIVE—pertaining to the mental process or faculty by which knowledge is acquired.

COMMAND—a directive specifying what is to be done, and how it is to be accomplished.

COMMEND—to express approval of; to praise.

COMPONENT—a part of a greater whole.

COMPREHENSION—an understanding of a thing or event.

CONCISE—expressing much in few words.

CONJECTURE—an inference or guess based on inconclusive or incomplete evidence.

CONNOTATION—that which is suggested or implied rather than literal.

CONSOLE—to make feel less sad; to comfort.

CONSTRUCTIVE CRITICISM—a judgment or evaluation delivered to an individual with the purpose of improving that individual or his or her performance.

CONTINGENCY—an event that might occur but that is not likely or intended; a possibility.

CORNERSTONE—the indispensable and fundamental basis of something.

CORRUPTION—the destroying or subversion of the honesty or integrity of something.

CREATIVITY—the ability or power to create or produce things.

CREDIBILITY—worthiness of belief in someone or something.

DEBACLE—a sudden disaster.

DEBILITATING—tending to make feeble or inadequate.

DECIPHER—to translate into ordinary language, make understandable, or make out the meaning of.

DECRY—to belittle or disparage openly; to depreciate or devalue.

DEFICIENCY—a shortcoming.

DEMEANING—debasing or degrading; tending to lower in dignity or status.

DIALOGUE—an exchange of ideas or opinions.

DISSOLUTE PRACTICES—actions lacking in moral restraint or good taste.

DOCILE—easily managed, trained, or taught.

EMPOWER—to enable, permit, or authorize.

ENTITY—something that has separate and distinct existence and objective or conceptual reality.

ENTOURAGE—a group of attendants, followers, or associates.

ENUNCIATE—to pronounce or articulate with clarity.

EQUANIMITY—the quality or characteristic of being even-tempered or well composed.

EQUILIBRIUM—mental or emotional balance; a condition of being stable.

ESPRIT DE CORPS—a spirit of enthusiasm among members of a group for one another, their group, and its purposes.

ESTHETIC—of or pertaining to the sense of the beautiful.

ETHICAL CODE—a principle of right or good conduct, or a body of such principles.

EXALTATION—the state or feeling of intense well-being or elevation in status.

EXPEDIENT—based on what is of use or advantage rather than what is right or just.

EXTERNAL DYNAMICS—the changing relationships between or among individuals comprising a group and individuals outside the group.

FEEDBACK—any information about the result of an action or process.

FERVENT—having or showing great intensity of interest or devotion.

FORTITUDE—the strength of mind that allows one to endure pain or adversity with courage.

GENERATIONS—offspring having a common parent or parents and constituting a single stage of descent.

THE "GRAPEVINE"—an informal means of transmitting information, gossip, or rumor from person to person within a group.

GROUP DYNAMICS—the factors and relationships that govern behavior of a group of people.

HALLMARK—any mark indicating quality or excellence, or any conspicuous indication of same.

HIERARCHY—a body of persons organized or classified according to rank, capacity, or authority.

HUMANE—having the good qualities of human beings, as kindness, mercy, or compassion.

ILLUSORY—deceptive through the creation of a false impression.

IMPERVIOUS—incapable of being affected or harmed.

INCOMPETENCE—not capable or able in some area of endeavor.

INCORRIGIBLE—incapable of being corrected or reformed.

INDOCTRINATE—to instruct in a body of doctrine, or system of thought.

INEPTITUDE—the quality of being not fitting or appropriate; awkward or clumsy.

INFLECTION—an alteration in pitch or tone of the voice.

INFRACTION—a violation or transgression.

INSIGHT—the ability to discern the true nature of a situation from other than direct knowledge.

INTANGIBLES—things that have immaterial as opposed to material value.

INTERNAL DYNAMICS—the changing nature of a relationship between or among individuals within a group.

INTERPERSONAL DYNAMICS—the changing nature of a relationship between or among people.

INTRIGUE—a secret or underhanded scheme to achieve some purpose.

INVIDIOUS—tending to arouse ill-will or animosity; discriminatory.

JUSTICE—fair handling; due reward or treatment.

LEGITIMATE—based on logical reasoning; authentic or genuine.

MALADJUSTMENT—the inability to adjust personality needs to the demands of the environment.

MALAISE—a sense of physical, mental, or moral ill-being or uneasiness.

MEDIOCRITY—the state of being only average or not particularly outstanding.

METAPHOR—a figure of speech, allegory, or parable.

MILESTONE—an important event or turning point in a process or career.

MULTIFACETED—having many sides or aspects.

MURPHY'S LAW—an axiom that states that if anything can go wrong, it will (usually at the worst possible time).

NEW DIMENSION—a new extent, size, or scope of something.

ONUS—responsibility or blame for a wrong.

ORDER—a directive specifying what is to be done, but not how.

OVERBEARING—dominating or arrogant.

PARALYSIS—the partial or complete loss of motion or feeling in some part of the body.

PERCEPTIONS—events or things that one becomes aware of through any of the senses.

PERFORMANCE REVIEW—a formal or informal assessment of an individual's actions during a set period of time.

PERNICIOUS HABIT—a repeated behavior causing moral injury or harm.

PERQUISITE OF COMMAND—a benefit of having command of a unit.

PERSPECTIVE—a specific point of view in understanding or judging things or events, especially one that shows them in their true relationship to one another.

PHYSIOLOGICAL—pertaining to the biological science of life processes, activities, and functions.

PONTIFICATE—to speak or behave with pompous authority.

PREDECESSOR—a person who precedes another; an ancestor.

PRESTIGE—prominence or influence achieved through success, renown, or wealth.

PRIMORDIAL—something original or fundamental.

PROCRASTINATION—the act of putting off doing something until a future time.

PRODUCTIVITY—the degree of production or yielding useful results.

PROPAGANDA—the systematic propagation of a given doctrine.

PROPENSITY—an inclination to do something.

PROXIMATE GOAL—a near-term objective, relatively easy to attain.

PRUDENCE—the exercise of wisdom in handling practical matters; good judgment.

PSYCHIC INCOME—something benefiting the ego or the mind.

PSYCHOLOGICAL FACTORS—factors pertaining to or derived from the mind or emotions.

PYRAMID STRUCTURE—an organization with many members at the lowest level and a diminishing number toward the top.

RATIONAL ARGUMENT—an argument or attempt to persuade based upon reason or logic.

RECONCILE—to settle or resolve a dispute, or to reestablish a friendship or beneficial relationship.

REQUISITE AUTHORITY—the needed or appropriate authority to do something.

RESONATE—to reinforce or prolong something, especially a sound, by sympathetic vibration corresponding to the original frequency.

SARCASTIC—sharply mocking or contemptuous.

SCAPEGOAT—a person or group made to bear blame for others.

SEMANTICS—the study or science of meaning in language.

SERVICE REPUTATION—the general estimation of the effectiveness or worth of a person working in a military environment.

SKEPTICAL—tending not to believe; doubting.

SLOTH—laziness or aversion to work or exertion.

SOCIAL FABRIC—the social environment in which a person or society operates.

SOCIOLOGIST—one who studies human social behavior.

SPAN OF CONTROL—the breadth of a process or number of individuals that a superior can effectively supervise.

STATUS—the position or standing, especially social standing, of an individual or group.

STATUS QUO—the existing condition or state of affairs.

STEREOTYPE—a preconceived conception, opinion, or belief, usually not conforming with the true nature of something.

STIMULUS—something that incites or rouses to action or causes a response.

STRAIGHTFORWARD—honest, frank, candid.

SUBSCRIBE—to express approval, assent, or agreement with.

SUBVERSIVE—pertaining to undermining the character, morale, or allegiance of a target group.

SUBVERT—to undermine the character, morale, or allegiance of an individual or group.

SYNDROME—a group of signs or symptoms that collectively indicate some abnormal condition.

SYNERGISTIC—describing a situation in which the total effect tends to be greater than the sum of the individual effects.

TECHNOCRATIC—tending to be controlled by science, scientists, or technicians.

TEMPERANCE—the exercise of moderation and self-restraint.

ULTIMATE GOAL—the final objective, which may be long term or difficult to attain.

UNAMBIGUOUS—not lacking in clarity; having only one meaning.

UNEQUIVOCAL—admitting of no doubt or misunderstanding.

VALUE SYSTEM—a group of principles or qualities that an individual considers worthwhile or desirable.

VITALITY—energy, or power of an individual or institution to endure or continue in existence.

VOCATION—an occupation or profession for which one is especially suited or qualified.

Bibliography

UNIT 1

Horton, Sgt. Maj. John L., USMC. "Look, Listen, and Lead." U.S. Naval Institute *Proceedings*, November 1987, pp. 84–87.

Montor, Karel, and Anthony J. Ciotti, eds. *Fundamentals of Naval Leadership*. 3d ed. Annapolis: Naval Institute Press, 1984.

Reserve Liaison and Training Branch, Education Center, Marine Corps Development and Education Command. *Leadership in Action: A Text for U.S. Marine Corps Junior ROTC*. Quantico, Va.: U.S. Marine Corps, 1974.

Trost, Adm. Carlisle A. H., USN. "Leadership Is Flesh and Blood." U.S. Naval Institute *Proceedings*, February 1988, pp. 78–81.

Washbush, Cdr. John B., USNR, and LCdr. Barbara J. Sherlock, USN, eds. *To Get the Job Done*. Annapolis: Naval Institute Press, 1981.

UNIT 2

Ballou, Sidney. "Faulty Communications." U.S. Naval Institute *Proceedings*, February 1929, pp. 89–98.

Bass, Bernard M. *Leadership, Psychology, and Organizational Behavior*. Westport, Conn.: Greenwood Press, 1960.

Bearden, Bill. *The Bluejacket's Manual*. 21st ed. Annapolis: Naval Institute Press, 1990.

Bekkehdal, C. L. "Discipline and the Profession of Naval Arms." U.S. Naval Institute *Proceedings*, May 1977, pp. 186–201.

Black, James M., and Virginia T. Black. *The Front-Line Manager's Problem-Solver*. New York: McGraw-Hill, 1967.

Boyd, Bradford B. *Management-Minded Supervision*. New York: McGraw-Hill, 1968.

Burke, Arleigh A. "Admiral Marc Mitscher: A Naval Aviator." U.S. Naval Institute *Proceedings*, April 1975, p. 53.

———. "The Art of Command." In *Leadership and Law Department NL 303 Book of Readings*. Ed. J. M. Kelly, G. D. Patton, J. F. Downs, and G. J. Corini. Annapolis: U.S. Naval Academy, academic year 1983–84, pp. 125–29.

Burton, John Wear. *Deviance, Terrorism, and War*. New York: St. Martin's Press, 1979.

Callahan, D., P. L. Stromberg, and M. M. Wakin. *The Teaching of Ethics in the Military*. Hastings-on-Hudson: The Institute of Society, Ethics, and the Life Sciences, 1982.

"Calley's Story." *U.S. News and World Report*, 8 December 1968, p. 4.

Carrington, James H. *Command Control Compromise*. Annapolis: Naval Institute Press, 1973.

Clexton, E. W., Jr. "Trust and Confidence." Commanding Officer's Memorandum no. 20, USS *Dwight D. Eisenhower* (CVN 69), 21 February 1984.

Collins, F. C. "The Loss of Leadership." U.S. Naval Institute *Proceedings*, April 1975, p. 53.

Dahl, Robert A. *Modern Political Analysis*. Englewood Cliffs, N.J.: Prentice-Hall, 1970.

Daniels, Anthony J. "Evaluation Systems and Superior-Subordinate Relationships." *Military Review*, January 1972, pp. 3–8.

Dean, B. C. "Authority: The Weakened Link." U.S. Naval Institute *Proceedings*, July 1971, pp. 48–52.

Diggins, John P., ed. *The Problem of Authority in America*. Philadelphia: Temple University Press, 1980.

Disciplinary Regulations of the Armed Forces of the USSR: The Officer's Handbook—a Soviet View. For sale by the Superintendent of Documents, Washington, D.C. Moscow, 1971, p. 167.

Dornbusch, Sanford M., and W. Richard Scott. *Evaluations and the Exercise of Authority*. San Francisco: Jossey-Bass, 1975.

Estes, Kenneth W. *The Marine Officer's Guide*. 6th ed. Annapolis: Naval Institute Press, 1995.

Gibson, James L., John M. Ivancevich, and James H. Donnelly, Jr. *Organizations*. 4th ed. Plano, Tex.: Business Publications, 1982.

Gillan, Richard, ed. *Power in Postwar America*. Boston: Little, Brown, 1971.

Hasson, Raymond E. "The New Breed of Sailor." U.S. Naval Institute *Proceedings*, March 1979, pp. 41–46.

Hazard, John G. Response to "Leadership." U.S. Naval Institute *Proceedings*, April 1981, p. 80.

Hellriegel, Don, and John W. Slocum, Jr. *Management*. 3d ed. Reading, Mass.: Addison-Wesley, 1982.

Helms, R. E. "Shipboard Drug Abuse." U.S. Naval Institute *Proceedings*, December 1975, pp. 41–45.

Holland, W. J. "Command at Sea." U.S. Naval Institute *Proceedings*, December 1976, p. 18.

Hollander, Edwin P. *Leadership Dynamics*. New York: Stein and Day, 1982.

Infield, Glenn R. *Secrets of the SS*. New York: Free Press, 1978.

Jacobsen, K. C. "The Stranger in the Crowd." U.S. Naval Institute *Proceedings*, September 1974, p. 33.

Janowitz, Morris. *The Professional Soldier: A Social and Political Portrait*. New York: Free Press, 1971.

Johnson, Spencer, and Kenneth Blanchard. *The One Minute Manager*. New York: William Morrow, 1982.

Kattar, R. J. "The First Commandment of Leadership: Love Thy Soldier." *Military Review*, July 1980, pp. 65–68.

Keenan, P. C., and C. W. Tarr. "People: The Navy's Most Critical Resource." U.S. Naval Institute *Proceedings*, November 1976, p. 45.

Knowels, Malcolm, and Hulda Knowels. *How to Develop Better Leaders*. New York: Association Press, 1955.

Krieger, Leonard, and Fritz Stern. *The Responsibility of Power*. Garden City, N.Y.: Doubleday, 1967.

Kuhne, Michael D. "Can Do! Should Do?" U.S. Naval Institute *Proceedings*, March 1977, p. 86.

Laffin, John. *Links of Leadership: Thirty Centuries of Military Command*. New York: Abelard-Schuman, 1970.

Lindgren, Henry Clay. *Leadership, Authority, and Power Sharing*. Malabar, Fla.: Robert Kreiger, 1982.

Loye, David. *The Leadership Passion*. San Francisco: Jossey-Bass, 1970.

Mack, William P. "The Need for a New Morality." U.S. Naval Institute *Proceedings*, December 1975, pp. 30–33.

"The Massacre at Song-My." *Life*, 5 December 1969.

Military Leadership: Supplementary Readings. West Point, N.Y.: U.S. Military Academy, Department of Tactics, Office of Military Psychology and Leadership, 1962.

Montgomery, Bernard. *The Path to Leadership*. New York: G. P. Putnam's Sons, 1961.

Montor, Karel, and Anthony J. Ciotti, eds. *Fundamentals of Naval Leadership*. 3d ed. Annapolis: Naval Institute Press, 1984.

Munson, Edwin Lyman, Jr. "Leadership for American Army Leaders." *Infantry Journal*, 1941.

"The My-Lai Massacre." *Newsweek*, 24 November 1969.

Noel, John V., Jr., and James Stavridis. *Division Officer's Guide*. 9th ed. Annapolis: Naval Institute Press, 1989.

O'Hara, M. J. "The Challenge of Moral Leadership." U.S. Naval Institute *Proceedings*, August 1977, pp. 58–62.

"An Old Sailor Tradition." U.S. Naval Institute *Proceedings*, April 1975, pp. 42–45.

Peers, William A. *The My-Lai Inquiry*. New York: Norton Books, 1974.

Powers, William H. "Almost beyond Human Endurance." U.S. Naval Institute *Proceedings*, December 1977, pp. 60–69.

Reed, B. *Personal Leadership for Combat Officers*. New York: Whittlesey House, 1943.

Reynolds, Clark G. "Youth and the U.S. Navy." U.S. Naval Institute *Proceedings*, July 1973, pp. 26–34.

Rocap, Pember W. "Who Was Tangled in the Chain When We Threw It Out the Window?" *Air University Review*, vol. 25, no. 4 (1974), pp. 75–80.

Roskill, S. W. *The Art of Leadership*. London: Collins Clear-Type Press, 1964.

Solley, George C. "Trust, Confidence, and Obligation." U.S. Naval Institute *Proceedings*, November 1981, pp. 72–74.

Soper, M. E. "Officers for the Eighties: A Challenge for NROTC." U.S. Naval Institute *Proceedings*, February 1975, p. 40.

Stavridis, James. "Closing the Gaps in Naval Leadership." U.S. Naval Institute *Proceedings*, July 1982, pp. 76–78.

———. "On Leading Snipes." U.S. Naval Institute *Proceedings*, January 1981, p. 74.

Stivers, E. R. "The Mystique of Command Presence." U.S. Naval Institute *Proceedings*, August 1968, pp. 26–33.

Vance, Charles C. *Boss Psychology*. New York: McGraw-Hill, 1975.

Wakin, Malham M. *Military Leadership: In Pursuit of Excellence*. Ed. Robert L. Taylor and William Rosenback. Boulder, Colo.: Westview Press, 1984, p. 55.

Walzer, Michael. "Two Kinds of Military Responsibility." *Proceedings of the War and Morality Symposium*. West Point, N.Y.: U.S. Military Academy, 1980, pp. 20–29.

Ware, Hugh C. "New Tools for Crisis Management." U.S. Naval Institute *Proceedings*, August 1974, p. 19.

Watkins, James D. "The Principle of Command." U.S. Naval Institute *Proceedings*, January 1983, pp. 32–33.

White, Jack M. "The Military and the Media." U.S. Naval Institute *Proceedings*, July 1974, pp. 47–51.

White, William S. *The Responsibles*. New York: Harper and Row, 1972.

Williamson, Porter B. *Patton's Principles*. Tucson: Management and Systems Consultants, 1979.

Wrong, Dennis H. *Power: Its Forms, Bases, and Uses*. New York: Harper and Row, 1979.

UNIT 3

Anderson, C. J. L. F. "The Defense of Superior Orders." *Russi Journal of the Royal United Services Institute for Defence Studies*, June 1981, pp. 52–57.

Argyris, Christopher. *Interpersonal Competence and Organizational Effectiveness*. Homewood, Ill.: Dorsey Press, 1962.

Arrison, J. M. "Battle Orders." U.S. Naval Institute *Proceedings*, July 1984, pp. 62–67.

Asprey, Robert B. *The Battle of the Marne*. Philadelphia: Lippincott, 1962.

Aurner, Robert Ray. *Effective Communication in Business*. Cincinnati: Southwestern, 1967.

Bach, C. A. "Leadership." In *Selected Readings in Leadership*. Ed. Malcolm E. Wolfe and F. J. Mulholland. 2d ed. Annapolis: Naval Institute Press, 1960.

Balkind, Jonathan J. *Morale Deterioration in the United States Military during the Vietnam Period*. Ph.D. diss., UCLA, 1978. Ann Arbor, Mich.: University Microfilms International, 1978.

Ballou, Sidney. "Faulty Communications." U.S. Naval Institute *Proceedings*, February 1929, pp. 89–98.

Becke, Archibald F. *Napoleon and Waterloo*. New York: Books for Libraries Press, 1971.

Bellows, Roger. *Creative Leadership*. Ed. Dale Yoder. Englewood Cliffs, N.J.: Prentice-Hall, 1959.

Brown, Leland. *Communicating Facts and Ideas in Business*. Englewood Cliffs, N.J.: Prentice-Hall, 1961.

Burrage, Henry. *Gettysburg and Lincoln*. New York: G. P. Putnam's Sons, 1906.

Cherrington, David J. *Personnel Management*. Dubuque, Iowa: William C. Brown, 1983.

Cherry, Colin. *On Human Communication*. Cambridge: MIT Press, 1966.

Clarke, Bruce C. *Guidelines for the Leader and Commander*. Harrisburg, Pa.: Stackpole, 1973.

Clarke, Jean Illsley. *Who, Me Lead a Group?* Minneapolis: Winston Press, 1984.

Connelly, J. Campbell. *A Manager's Guide to Speaking and Listening*. New York: American Management Association, 1967.

Department of the Navy. *Quotable Navy Quotes*. Washington, D.C.: United States Office of Information, 1967.

"Effective Communications." *Approach*, August 1971, p. 3.

Eisenhower, Dwight D. *At Ease: Stories I Tell to Friends*. Garden City, N.Y.: Doubleday, 1967, p. 323.

Estes, Kenneth W. *The Marine Officer's Guide*. 6th ed. Annapolis: Naval Institute Press, 1995.

Faris, John H. "Leadership and Enlisted Attitudes." In *Military Leadership*. Ed. James H. Buck and Lawrence J. Korb. Beverly Hills: Sage, 1981.

Fogel, Irving M., and Anne E. Bell. "Communicating Effectively in the Construction Industry." *Military Engineer*, September–October 1981, pp. 344–46.

Freeman, Douglas S. "Leadership." In *Selected Readings in Leadership*. Ed. Malcolm E. Wolfe and F. J. Mulholland. 2d ed. Annapolis: Naval Institute Press, 1960.

Furse, George Armand. *Information in War: Its Acquisition and Transmission*. London: William Clones and Sons, 1895.

Galdston, Iago, and Hans Zetterberg. *Panic and Morale*. New York: International Universities Press, 1958.

Gibson, James L., John M. Ivancevich, and James H. Donnelly, Jr. *Organizations*. 4th ed. Plano, Tex.: Business Publications, 1982.

Goodwin, J. F. "Ship's Orders—1815." U.S. Naval Institute *Proceedings*, January 1940, pp. 78–82.

Gorden, Thomas. *Leadership Effectiveness Training*. New York: Wyden Books, 1977.

Hamilton, Andrew. "Where Is Task Force Thirty-Four?" U.S. Naval Institute *Proceedings*, October 1960, pp. 76–80.

Hawkins, Wallace W. "Participate." U.S. Naval Institute *Proceedings*, August 1966, p. 91.

Hays, Samuel H., and William N. Thomas. *Taking Command*. Harrisburg, Pa.: Stackpole, 1967.

Heinl, Robert Debs, Jr. *Dictionary of Military and Naval Quotations*. Annapolis: Naval Institute Press, 1966.

Hellriegel, Don, and John W. Slocum, Jr. *Management*. 3d ed. Reading, Mass.: Addison-Wesley, 1982.

Henderson, George, ed. *Human Relations in the Military*. Chicago: Nelson-Hall, 1975.

Hubbell, John G. *POW*. New York: Reader's Digest Press, 1976.

Hudgins, L. E. "Ready for Fear." In *Selected Readings in Leadership*. Ed. Malcolm E. Wolfe and F. J. Mulholland. 2d ed. Annapolis: Naval Institute Press, 1960.

Hurlburt, J. S. Response to "Communications and Command Prerogative." U.S. Naval Institute *Proceedings*, July 1974, p. 96.

Jones, William. "Getting Things Done through Better Communications." *U.S. Army Aviation Digest*, October 1974, p. 35.

Jurika, Stephen, Jr. *From Pearl Harbor to Vietnam*. Stanford, Calif.: Hoover Press, 1980.

Kaufman, Herbert. *Administrative Feedback*. Washington, D.C.: Brookings Institution, 1973.

Keena, E. Douglas. "Exploring Leadership." *Vital Speeches of the Day*, August 1980, pp. 638–40.

Lawrence, James J. "Effective Communication." *Aerospace Safety*, December 1977, pp. 2–5.

Leighton, A. H. *The Governing of Men*. Princeton, N.J.: Princeton University Press, 1946, p. 288.

Linver, Sandy. *Speak Easy*. New York: Summit Books, 1958.

McCracken, Alan. "When Is an Order?" U.S. Naval Institute *Proceedings*, June 1949, pp. 653–57.

Mack, William P., and Albert H. Konetzni, Jr. *Command at Sea*. 4th ed. Annapolis: Naval Institute Press, 1982.

Mack, William P., and Thomas D. Paulsen. *The Naval Officer's Guide*. 10th ed. Annapolis: Naval Institute Press, 1991.

"Manager Communications: Organization Lifeline." *Defense Management Journal*, January 1973.

Markham, Felix. *Napoleon and the Awakening of Europe*. New York: Collier Books, 1965.

Marshall, D. J. "Communications and Command Prerogative." U.S. Naval Institute *Proceedings*, January 1974, pp. 29–33.

Megginson, Leon. *Personnel: A Behavioral Approach to Administration*. Homewood, Ill.: Richard D. Irwin, 1967.

Mooney, J. D. *Principles of Organization*. Rev. ed. New York: Harper Brothers, 1947, p. 131.

Newman, Aubrey S. *Follow Me: The Human Element in Leadership*. Novato, Calif.: Presidio Press, 1981.

Newman, William H. *Administrative Action*. 2d ed. Englewood Cliffs, N.J.: Prentice-Hall, 1963.

Plumb, Charlie, and Glen DeWerff. *I'm No Hero*. Independence, Mo.: Independence Press, 1973.

Potter, E. B. *Bill Halsey*. Annapolis: Naval Institute Press, 1985, p. 290.

———, ed. *Sea Power: A Naval History*. 2d ed. Annapolis: Naval Institute Press, 1981.

Puryear, Edgar F. *Nineteen Stars*. Washington, D.C.: Coiner, 1971.

"Realities of Clear Communications." *Army Logistician*, March–April 1977.

Redfield, Charles E. *Communication in Management*. Chicago: University of Chicago Press, 1958.

Roskill, S. W. *The Art of Leadership*. London: Collins Clear-Type Press, 1964.

Samovar, Larry A., and Jack Mills. *Oral Communication*. Dubuque, Iowa: William C. Brown, 1972.

Selby, John. *The Thin Red Line at Balaclava*. London: Hamish Hamilton, 1970.

Ship Organization and Personnel. Annapolis: Naval Institute Press, 1972.

Smith, S. E. *The United States Navy during World War II*. New York: William Morrow, 1966.

Soper, Paul L. *Basic Public Speaking*. New York: Oxford University Press, 1949.

Staff, W. W. "Mission-Type Orders." *Army*, January 1957, p. 71.

Stavridis, James. *Watch Officer's Guide*. 13th ed. Annapolis: Naval Institute Press, 1992.

Thayer, Lee O. *Administrative Communication*. Homewood, Ill.: Richard D. Irwin, 1961.

Vaeth, J. Gordon. "Communications: A Lost Art?" *U.S. Naval Institute Proceedings*, September 1976, p. 80.

Valenti, Jack. *Speak Up with Confidence*. New York: William Morrow, 1982.

Van Dersal, William. *The Successful Supervisor*. New York: Harper and Row, 1974.

Vardaman, George T. *Effective Communication Ideas*. New York: Van Nostrand Reinhold, 1970.

Vardaman, George T., and Carroll C. Halterman. *Managerial Control through Communication*. New York: John Wiley and Sons, 1968.

Wakin, Malham M. "Ethics of Leadership." In *Military Leadership*. Ed. James H. Buck and Lawrence J. Korb. Beverly Hills: Sage, 1981.

Weinhold, Barry K. *Transpersonal Communication*. Englewood Cliffs, N.J.: Prentice-Hall, 1979.

Woodham Smith, Cecil Blanche. *The Reason Why*. New York: E. P. Dutton, 1960.